BAR SERVICE

BAR SERVICE

James H. Coombs

with a foreword by
J. G. MILES, M.I.P.M., F.I.T.O.
Secretary, Licensed Trade Training
and Education Committee

drawings by
BILL HOOPER

Hutchinson

London Melbourne Sydney Auckland Johannesburg

Hutchinson & Co. (Publishers) Ltd

An imprint of the Hutchinson Publishing Group

17–21 Conway Street, London W1P 6JD

Hutchinson Group (Australia) Pty Ltd
30–32 Cremorne Street, Richmond South, Victoria 3121
PO Box 151, Broadway, New South Wales 2007

Hutchinson Group (NZ) Ltd
32–34 View Road, PO Box 40–086, Glenfield, Auckland 10

Hutchinson Group (SA) (Pty) Ltd
PO Box 337, Bergvlei 2012, South Africa

First published by Barrie and Rockliff 1965
Second edition published by Barrie & Jenkins 1975
First published by Hutchinson 1982
Reprinted 1983

Printed in Great Britain by The Anchor Press Ltd
and bound by Wm Brendon & Son Ltd
both of Tiptree, Essex

ISBN 0 09 147541 4

CONTENTS

ACKNOWLEDGEMENTS

Acknowledgement and thanks are given to the following firms and their officers for the kind assistance so freely given in the compilation of this work: Arthur Guinness Son & Co. Ltd., Bass, Ratcliffe & Gretton Ltd., Ind Coope Ltd., Watney-Mann Ltd., Courage Barclay Ltd., Whitbread & Co. Ltd., Benskins Watford Brewery, Carlsberg Distributors Ltd., Holsten Distributors Ltd., Pilsner-Urquell Co. Ltd., R. N. Coate & Co. Ltd., Moussec Ltd., Vine Products Ltd., Gonzalez Byass Ltd., Gilbey Twiss Ltd., Findlater, Mackie Todd & Co. Ltd., Harveys of Bristol, Jarvis, Halliday & Co. Ltd., Williams and Humbert Ltd., Mr. Schlach (on Bottles), Harry Shimeild & Sons Ltd., John Player & Sons Ltd., The proprietors and staff of *The Morning Advertiser*, J. C. McLaughlin Ltd., Tate & Lyle Ltd., United Rum Merchants, Booth's Distilleries Ltd., Matthew Clark & Sons Ltd., Finsbury Distillery Co. Ltd., Pimms Ltd., Luis Gordon & Sons Ltd., Evans, Marshall & Co. Ltd., J. R. Parkington Ltd., Sangers Ltd., Gaskell & Chambers Ltd., Wm. Page & Co. Ltd., G. B. Hobson & Son Ltd., Smith's Potato Crisps Ltd., Gallagher Ltd., The proprietor and staff of *The Daily Telegraph*. Thanks are also due to the many other individuals who have offered the hints, suggestions and advice which have enabled me to make the book as comprehensive as possible.

The author is particularly grateful to Mr. Peer Groves, formerly Chairman of the Licensed Trade Education Committee, who read the whole work in manuscript, making many useful suggestions and constructive criticisms.

Publisher's note. This second edition was revised throughout by the author shortly before his death in 1975.

FOREWORD

Mr. James Coombs wrote this book for learner staff and some of the information given may seem extremely elementary. In some cases the advice given may be found to cut across long-established practice, but where this occurs the reason behind it is also given. However elementary or however revolutionary it may seem, this information is part of the knowledge essential to a barman or barmaid who wishes to have pride in a job well done.

In the foreword to the first edition, the late Mr. J. W. Peer Groves mentioned Mr. Coombs' insight and description of the foibles of customers, but endorsed his view that a barman must recognise that customers' idiosyncrasies are their own business—provided they do not cause discomfort to other customers.

The author was fully conscious of the fact that there is an emphasis throughout on Southern England, where his own not inconsiderable experience was. Readers in other parts of the British Isles must therefore be prepared to make their own mental adjustments when they know that local usage, traditions, tastes and vocabulary differ. The principles hold good everywhere. As Mr. Peer Groves said, it may be read by any beginner with benefit, by those of greater experience with profit and by all, whatever their job in the trade, with pleasure.

If many drinks nationally known and some more local are referred to by name, this was because Mr. Coombs himself decided it was necessary to describe in detail their correct service. He apologised that other brands of equally commendable merit received no specific mention. This caused him considerable heart-searching, but short of writing in the full range of every single supplier, the possibility that some might feel slighted had to be accepted. The criterion in every case, he assured readers, was the value of the information imparted by the naming of a representative selection of goods. Consolation might, he felt, be drawn from the fact that where such an omission is noted by the reader, the object of the book will have been achieved in establishing a point of comparison between the products mentioned and those left out. If a barman affirms the merits of product "X" someone on the other side of the bar will mention product "Z" and thus increase trade education to the ultimate benefit of all.

This second edition differs from the first in that it has been updated, but with no addition to its predecessor. It never pretended to be a complete training manual, and it seemed best to the publishers to leave it as it was, except where it has been overtaken by events, and so retain its original style and flavour.

J. G. MILES, M.I.P.M., F.I.T.O.,
Secretary, Licensed Trade Training &
Education Committee.

INTRODUCTION

So you fancy entering the Licensed Trade? You have thought it over and made up your mind that serving drinks to an unappreciative and sometimes downright rude public is just the life for you?

At the present moment you have no experience, know nothing of what goes on behind the scenes, think all looks very easy going, comfortable, warm and happy, can't tell a White Shield from a White Lady, but nevertheless, think you are going to like it.

To make a real success of Barmanship *you have got to like it*. That is the first requisite. It is no use trying to fool anyone about the Trade in an endeavour to entice them into it, for the young man or woman who enters it grudgingly is doomed to failure from the start.

Sort yourself out under the following headings and see if you are—sweet-tempered, long-suffering, patient, polite, healthy, hard-working, willing to learn—and, above all, honest. Be *honest* with your answers, for every one of these attributes (and a few more) is essential to you in your new career.

The Trade offers some very fine opportunities to young people, with good pay and comfortable living, comparable to anything available outside, provided every effort is made to learn it properly from the start, and provided they "play the game!"

From the customers' side of the counter some very strange ideas prevail about the "wonderful life" behind the bar. These often stem from semi-alcoholics who think it must be heaven to be surrounded by unlimited drink. Never, for one moment, do they give a thought to what happens after "permitted hours". They stagger home at night leaving the bar a complete shambles, in the shambling of which they gave ample assistance, and enter again in the morning to find everything spick and span. The counters and furniture are polished, the floors, carpets and brass cleaned, the fire going nicely, the glasses shining, the shelves sparkling, the bottles all lined up and the beer all ready to serve. All this by magic! Don't you believe it—you'll have done your share towards it.

Do not let this picture put you off. You, being young, enthusiastic and energetic, will take it all in your stride. You will take a pride in your cabinet and feel a sense of guilt if one speck of dust can be found in that part for which you are held responsible. Once you know what is in each and every bottle, how it has been made and how it should be served your interest will be doubled.

The information given in these pages is going to help you tremendously in your job—especially at the beginning. It will help to save you making a fool of yourself, or anyone else making a fool of you. Study it carefully—don't just skim through it. Make notes if you feel they will help, and read every other piece of printed matter you can find concerning your wares—leaflets, booklets, labels—attend courses if you can. You will be surprised how some people will turn to you for information once they realise that you know the answers.

From here onwards this book can open the door to a new life for you—the gateway to a career which could lead to your being well established in life, with few of the worries which beset the outside world.

JAMES H. COOMBS
London, March 1975

PART I
BEER, CIDER, PERRY, SOFT DRINKS AND MIXERS

BEER—INTRODUCTORY

Over the years there has been a fluctuating tendency between the sales of bottled beer and draught beer (draught beer is that which is dispensed from pumps or taps).

Draught beers fell into a decline for various reasons including:

(a) Licensees allowing inexperienced staff to handle draught beers in their cellars—to the detriment of the Trade generally. Experienced cellarmen have always been much sought after and many publicans have been obliged to make do with second best.

(b) Governments whose increasing taxation had in many cases forced decreasing gravities. The duty and VAT on an average pint of beer selling at 22p at time of writing totals 7½p.

(c) Changing public taste coupled with a vastly increased purchasing power in the hands of ordinary people. It was usual before World War II for the ordinary working man to come alone into the public bar for half a pint of ale and five "Weights" (total 4½d). Now he may come into the saloon with his wife or daughter for a lager and lime and a large gin and tonic and twenty "Senior Service"—and good luck to him!

The days when a dozen hogsheads[1] of ale were seen being delivered to the corner pub have long since passed, but the Brewers, aware of current trends, developed Pressure Beer[2] in the shape of "Keg" or "Canister" beer, which has maintained and even strengthened the premier position which draught beer has always held.

It is important that you should make yourself familiar with the full range of beers on sale in your bar as soon as

---

[1] Hogshead—Cask containing 54 gallons.
[2] Pressure Beer—see p. 23.

1

possible—and remember the terms commonly used for them. Customers will ask you for a "Baby Ben", a "Mackey", a "D.D.", a "J.C." or a "Red" and you will not look very intelligent if you have to enquire what they mean.

Remember that whatever your range and sale of other drinks, most regular customers will judge a pub on how it keeps and serves its draught beer.

1
BOTTLED BEER AND ITS SERVICE

PALE ALE
There will always be a Pale Ale on sale. This is more often referred to as "Light" or "Light Ale".

The description "light" means light in alcoholic strength and does *not* refer to the colour. Opposite to the "light" are the "heavy" and in some parts of the country certain beers are ordered as "heavy"—meaning strong.

On occasions a brewer is found who labels his Pale Ale as "Pale Light Ale" or "Light Pale Ale", according with ancient usage. The labelling of beer is currently under review.

BROWN ALE
Brown Ale is so called on account of its colour—but it is, nevertheless, a "light" ale—i.e. alcoholically light. Generally speaking Pale Ale and Brown Ale represent the bulk of the bottled beer trade in most London and suburban houses— apart from Guinness.

SPECIAL BEERS
Next to the Pale and Brown Ales come the Special Beers, heavier alcoholically and well advertised by the national brewers. Ind Coope have their "Double Diamond" (D.D.), Courage their "John Courage" (J.C.), Truman their Ben Truman ("Ben")—and so on.

They are, of course, dearer than the light beers and are the pride of their respective breweries. They are pale in colour and sparkling. Rarely, if ever, does anything go wrong with any of these beers—it would be an event to find one of them out of condition except through negligence.

These beers are usually pasteurised. The purpose of pasteurisation is mainly to kill off any yeast cells remaining in the beer after it has been filtered as these may develop and cause a further undesired fermentation. One effect of

3

this process is to give these beers a longer shelf life.

The word "Export" on the label usually means that the beer has been pasteurised.

HEAVY BEERS

There will also be heavy dark beer on sale—usually bottled in "nips" ($\frac{1}{3}$) and known by the names given to them by their individual brewers, e.g. "Barley Wine", "Stingo",[1] "Arctic Ale", "Royal Toby", etc.

INDIA PALE ALE (I.P.A.)

This is a pale bitter beer sold in bottle and on draught. It is generally known as I.P.A.

The accepted reason for it being named "India" Pale Ale is that it used to be shipped to the East India Company who demanded very high standards and consequently the beer became known as a high quality pale ale, the name then being adopted for quality beers in this country.

Whitbreads have evidence of shipping beer to India from 1754 until the end of the East India Company's rule. A great demand for this was, of course, created by the Army in India.

There is a story that beer in bulk was used as ballast for sailing ships going out to India and on return to England was used for the home market. This formed a useful double purpose, both to the shipping company, and also to the brewers in establishing the fact that the beer would stand up to a rough passage, and also long delay in cask. This lacks confirmation, but it is thought to be fairly authentic.

What is undoubtedly a fact, however, is that a shipment of Pale Ale consigned from Bass's Burton brewery to India in 1827 was wrecked in the Irish Channel and part of it, on being salvaged, was sold in Liverpool for the benefit of the underwriters. The quality was so much appreciated that the fame of India Pale Ale was soon known throughout the country.

COLNE SPRING ALE

Wherever men foregather and regale themselves over great feats of drinking sooner or later the name Colne Spring

---

[1] Spices, pepper and other stimulants were commonly added to ale in former times in order to provide a bite—thus "Stingo". "Mulled" ale, i.e. ale with sugar and spice added and then heated, was also a cold-weather favourite.

is bound to crop up.

This was a strong and very potent ale, and if you hear any man boast that he drank ten pints in one evening and then walked home you can safely say he is not telling the whole truth—he probably walked home two days later!

When the Brewers—Benskins of Watford—were absorbed by Ind Coope, later to be absorbed within Allied Breweries, it was obvious that many beers had to be rationalised. Colne Spring suffered the treatment. First it became "Treble Gold" and is now bottled as a barley wine under the label "Triple A". It is sold in nip sizes and is very potent.

SEDIMENT BEER

The best known of the sediment beers is Worthington White Shield. This is allowed to mature in bottle and this process gives the beer its natural sparkle through fermentation. It also causes the beer to throw a sediment which acts as a preservative and this may be clearly seen at the bottom of each bottle. This sediment should be disturbed as little as possible after the beer has reached your shelves.

Whilst this sediment is not in any way harmful if poured into the glass it will obviously make the beer look very cloudy and unpalatable. Even so you may find some customers who insist on having the sediment poured into the glass—sometimes pouring it in themselves. When the bartender has taken every care to pour these beers so that they sparkle like purest crystal it is a trifle disconcerting to have a customer ruin their appearance with the bottom of the bottle. However, everyone to their taste!

POURING AND SERVING SEDIMENT BEERS

Great care must be taken in the handling of sediment beer. Bottles must be lifted carefully off the shelf, opened without jerking, and poured with a steady even flow against a good light at eye level so that the sediment can be observed.

When the sediment starts to move towards the neck of the bottle it is, of course, time to stop pouring, leaving the rest in the bottle. You should then have a glass of sparkling amber fit for a king!

Since, however, there *are* the eccentrics who enjoy "The Bottoms" as a final "Liqueur"—and your job is to please the clients—it is sometimes advisable to leave the bottle in front of the customer.

Should you be unfortunate enough to knock a bottle over

do not serve it. Place it on a shelf—apart from the others—
and leave it for a few hours. (If it has had a bad shaking it
may take twelve hours before it is sufficiently settled.)

Always see that new stock brought from the cellar is kept
separate from the old settled stock on your shelves (the
cellarman should see to this) and *always* serve in strict
rotation.

It is interesting to note that a bottle of White Shield
Worthington is considered by many to be the finest "pick-
me-up" in the morning after a "heavy" night out—and there
seems much to justify the claim.

LAGER

Of recent years the sales of lager beers have soared annually and represent a phenomenal increase over pre-war
years. For this reason it is essential for bar staff to know
something about this side of the trade.

The word "lager" is German for "store" or "storehouse".
Lager beer is therefore a beer which has been stored, i.e.
it has been stored to mature it. The slow fermentation at
low temperatures and the long maturing process give lager
an almost indefinite shelf life. A great favourite with the
German people, it was first made at Pilsen in Czecho-
slovakia by Urquell (brewers in Pilsen since 1295!)

The Pilsner Urquell Company take exception to other
brewers designating their lager "Pilsner" or "Pilsener",
claiming that only beer brewed in the town of Pilsen is
properly entitled to be thus named. The brewers, on the
other hand, claim that the word "Pilsner" or "Pilsener"
indicates merely a *method* of brewing lager. Some have
dropped the word from their labels, others still retain it—
and there is nothing the Pilsner Urquell Company can do
about it:

Apart from Urquell many Continental lagers are on sale
in this country, perhaps the best known being Carlsberg
and Tuborg from Denmark, Holsten and Lowenbrau from
Germany, Orangeboom and Heinekens from Holland, and
many others. Some of these are now brewed in England.

Holsten, brewed in Hamburg, deserves a wider market
than it receives here, being a first-class brew, stored
(lagered) for six months prior to shipment—and is a natural
lager (not carbonated) as is the case with beers not fermented and matured by the slow process.

Carlsberg deserves special mention as one of the foremost

6

Continental lagers. The vastness of the Copenhagen brewery (visited by Queen Elizabeth II and Prince Philip in 1957) is shown by the fact that the "lagers" hold the equivalent of 150 million bottles of beer. The Carlsberg Foundation is quite unique in brewing history as the total profit is given towards scientific and charitable interests.

British breweries have stepped up their output in the last few years, both bottled and draught.

"Harp" lager, very pale and pleasant and backed by massive advertising, is one well-known brand.

It is a fact that very few hardened beer drinkers entertain lager. It is perhaps best designated as a slightly "off-beat" drink with a certain snob appeal—eminently suitable for ladies who enjoy a little beer occasionally. Its growing popularity, however, even amongst men, stems from the fact that it must be served cold. There is a virtual guarantee that when a lager is called for it *will* be served cold—which gives it an appeal to the very thirsty on a hot day and others who enjoy a cool drink at any time—summer or winter.

The Americans started a vogue by adding Lime Juice Cordial to lager which give it a "fresh" tang and this practice has now become commonplace, for probably as much Lager and Lime is sold today as straight lager. Younger customers sometimes call for Lime Juice in Pale Ale—presumably for the same effect.

STOUT

Brewed from malts which have been well roasted (accounting for the darker colour) they are on the market under many descriptions. Some brewers produce several stouts of different characteristics—varying from sweet to dry. There are many excellent stouts and you will soon discover the most popular brands in your own locality and your own bar, but two deserve your special attention—because both are nationally advertised and live well above their advertised reputations. They are Guinness and Mackeson.

Mackeson

Originally this was named Mackeson's Milk Stout. It is thought to have been the first Milk Stout and was brewed at Mackeson's Brewery in Hythe, Kent, as far back as 1907 —hence the label reading "The original and genuine". The sale of this stout was not too brisk, but in 1929 Whitbreads

7

PICK·ME·UP

acquired Mackeson's Brewery and started to increase the distribution of Mackeson's Milk Stout. This created a market which attracted other brewers and they commenced to produce their own Milk Stouts.

Whitbreads, due to this competition, felt that the title "Milk Stout" might hamper the sales of Mackeson, and in 1946 the word "Milk" was dropped. This decision was followed by a Ministry of Food ruling which necessitated the dropping of the word "Milk" from *all* milk stouts, hence it became known as Mackeson's *stout* until 1961 when it was abbreviated to its present form of "Mackeson". So popular has it become that it is now abbreviated even further—from Mackeson to "Mackey" and from "Mackey" to "Mac".

Milky very properly describes Mackeson, which is a smooth, sweet, creamy stout and is deservedly popular with those people who do not appreciate the "bite" and vigour of the dry stout.

Guinness

Guinness is the most famous of all stouts; it was first brewed in Dublin in 1759 and is now brewed in nineteen countries throughout the world and is distributed to a hundred and forty countries.

In retail outlets Guinness should be stored at a temperature of betwen 55° and 62°F. to ensure that the Guinness is always in perfect condition, and of course you should always watch your stock and make sure that there is a strict rotation in the cellar or stock room. Generally speaking, you should keep at least a week's stock in hand, for bottled Guinness is best consumed when it is something like twenty-one to twenty-eight days from the time it leaves the Brewery.

So far as pouring bottled Guinness is concerned it is quite a simple operation if it is done carefully. You should never hold a glass upright and just tip the bottle and pour it in. If you do that you will probably get a lot of froth. What you need to do is hold the glass at an angle and then pour the Guinness in gently down the side of the glass and you can then see how the head is nicely developing, and as the glass becomes fairly full bring it to the upright position and you will find that you have got a lovely creamy head. Just to go back on that one—never hold the glass upright and just tip the Guinness in.

Barclay's Russian Stout

This beer was first brewed in the 1780's to the special order of the Russian Imperial Court and was designed to be especially strong to stand up to the rigours of a Russian winter. Its strength and flavour soon made it popular at home, too, and it has been brewed unchanged ever since.

It is a very strong, all-malt stout, and it goes through a three-year period of conditioning in cask and bottle. It is unusual among British brews in that it is a "vintage" brew —the year of the brewing being shown on each bottle. It is sold in nips ($\frac{1}{3}$ pint) and is necessarily expensive.

There are, of course, many other excellent stouts not mentioned above, but those mentioned are the widest known.

POURING AND SERVING BOTTLED BEERS

The pouring of *all* bottled beer should be done in front of the customer, at eye level, label facing the customer, as though you are proud to serve him with that very valued drink. A little bit of flourish is better than anything which savours of secrecy. For this reason never pour any beer at hip level, or on the pewter, as this practice gives rise to doubts and suspicions (unfounded in your case, of course!) and often calls forth caustic comments from the customers —some of which may not be over-polite!

When pouring bottled beer the most important thing is the *glass*. Make absolutely certain you are about to use the correct one and that it is not only visibly clean, but absolutely sparkling. If there is any doubt whatever about it (and even if there isn't) give it an extra polish with your clean glass cloth. Nice people appreciate good service.

Always select your glass before removing the crown cork from the bottle for two reasons. One: The beer may be high and some may be lost while you set it down and look for a glass. Two: It looks slightly ridiculous if, after having opened a bottle, you have to start searching for a glass.

Tilt the glass so that it is nearly level and start to pour the beer down the side WITHOUT ALLOWING THE BOTTLE NECK TO ACTUALLY TOUCH THE GLASS. Watch the beer carefully as you pour: if it commences to show a froth— slow down. If it looks as though you are going to get *nothing* more than a glass of froth—stop pouring altogether, stand the glass on the counter for a few seconds and wait for it to quieten down—then start pouring again—very slowly.

Sometimes it happens that a whole shelf of bottled beer

LET HIM SEE
WHAT YOU'RE DOING

11

pours out in lively fashion, possibly due to a shaking-up, or some other reason. In this case *don't* jerk off the crown cork—ease it off very gently and then keep the bottle on the tilt for a second or two. This is to allow the gas to escape and will make pouring much easier.

NEVER allow the beer to run over the top of the glass, especially a "stem" or footed one. The beer will come to rest on the foot of the glass and is likely to drip over clothing, etc. as soon as the customer commences to drink. With a little judgment this can be avoided, although it may sometimes mean leaving the bottle beside the glass until the rest of the beer can be poured. If you do make a mistake and happen to see beer running down to the foot of the glass wipe it with a clean glass-cloth immediately—before the customer picks it up—and make the necessary apologies!

It goes without saying that you throw away any glass which happens to be chipped or cracked *immediately* in case by any mischance it is used again. It is better to smash every faulty glass to fragments as a safeguard.

In serving all bottled beer NEVER, ON ANY ACCOUNT, ALLOW THE NECK OF THE BOTTLE TO COME IN CONTACT WITH THE BEER (or with the rim of the glass). This is a filthy habit, bearing in mind the dirt (and germs) which accumulate on bottles in the brewhouse yard, the

— EYE THIS FILTHY HABIT
WITH DISFAVOUR

journey on the dray and in your own cellar. Regrettably, it
is quite a common practice among flash barmen who pour
out two beers at once and dip both bottle necks well down
in the beer. This serves no useful purpose in any event and
most of your customers will eye this habit with distinct
disfavour, if not positive alarm.

It may shortly become law for *every* drink to be served in
a fresh glass, no matter how many times the customer returns his glass to the bar.

To some it may appear that this is carrying hygiene a
little too far, but a little thought will show that this is only
commonsense. However, it is likely to antagonise certain of
those who insist on being served with the same glass each
time they call. It is virtually impossible to avoid switching
and mixing up glasses when more than one is handed over
for re-filling and the spread of disease is therefore made
easy—for no man can say with any assurance that his com-
panion (be it even his own brother!) is not suffering from
some hidden complaint, or is a carrier. In the case of the

13

customer who insists on the SAME glass or glasses, it would be better to wash and dry each first, but time often does not permit of this.

FLAT BEER

A tip worth remembering in the case of beer which pours out "flat", due possibly to its being cold, is to take a glass, dip it in hot water, dry it, and pour out the beer vigorously into the still warm glass. This is often effective in producing a "head". There are, however, several reasons why beer pours out flat, either bottled or draught:

(1) Damp Glasses

Beer served in wet glasses often falls flat. When faced with a complaint of this nature see if any other beer around is also flat. If it appears to be just an isolated case (i.e. there is nothing seriously wrong with the beer itself) take a perfectly dry glass and pour the flat beer over into it briskly when it will probably show up a nice head—perhaps too much!

(2) Grease

Ascertain (being careful of course not to give offence) if the customer has been EATING anything of a greasy nature—crisps, bread and cheese, pies, etc. Even a small amount of grease on the lips is enough to send any beer flat. There is no remedy for this as nothing can be done to replace the head on a glass of greasy beer.

(3) Washing-Up

Excessive detergent in the washing-up water, when through pressure of service, glasses are drained and not dried or polished, has the same effect as grease. Only a very small amount of detergent is required in which to wash your glasses and the directions as to quantity must be carefully followed. The water should be very warm (not boiling) and the detergent added *after* it is drawn—*not* the other way round. The washing-up water should not show a mass of soapy bubbles on top and adding the detergent to the water prevents this. No soapy bubbles should be found in the glasses after they have been dipped and must certainly not be allowed to remain there. If any are visible then the glasses must be first rinsed in clean cold water before being dried.

(4) Hand Cream

Barrier cream on a barmaid's hands, often unsuspected and probably undetected, will also be enough to affect the head on beer if she is engaged in washing-up. Excessive detergent in the water IS inclined to make the hands rough, but the correct amount will barely affect them. The best remedy for chapped or rough hands is the regular use of "Glymiel" well rubbed in at bedtime—not, of course, during service.

Whilst on the subject of washing-up the following points must be borne in mind:

The water should be changed frequently and fresh detergent and sterilant added.

Counter swabs must NEVER be rinsed in the same water as is being used for washing-up.

The hands should not be washed in the well (or dried on a glass cloth). Retire to the wash-room for this purpose. In any properly run house there should be a hand towel behind the bar for drying the hands after washing-up.

If a glass happens to get broken in the well make absolutely certain that *every* piece is picked out to avoid accidents.

Treat glasses gently when washing-up. Cast-iron glasses have not yet been invented and the ordinary type do not stand up to rough handling. Be careful not to catch them on the edge of the tap or on the edge of the well; these are both common causes of breakage.

Water is cheap—use plenty of it

Regulations concerning hygiene in Public-houses, many of which were long overdue, are beginning to reflect in the extra care now being taken in serving food and drink. The public themselves are not far behind in demanding a better standard than was formerly thought necessary.

When serving drink of any kind do not hold the glass at the top. A moment's thought will show that the customer's lips will come into direct contact with your finger-prints, which most likely are well contaminated by germs from dirty swabs, bottles or other people's glasses.

Always handle glasses at the bottom

When serving any drink in a stem glass, or goblet, hold it in the cupped hand with the stem between the fingers.

15

Never, on any account, pick up any glasses for service by trying to gather up four in one hand—with a finger inside each. This might earn you a rebuke from a self-respecting navvy; it certainly will from a person of fastidious up-bringing. The next time you happen to be in a first-class hotel watch how the expert wine-waiter handles his glass-ware. In general it will always pay you to watch at work those with greater experience and skill than you have yet been able to acquire.

Should you be called upon to serve a tray of drinks at table always carry a clean glass-cloth, or napkin, over your arm—not under your armpit! This will be used for wiping up any accidental spillage. Always take clean drip-mats and place the glasses on these, removing any soiled ones. Remove, empty and clean up the ash-tray at the same time.

Serve the ladies first, and from behind, if this is possible and not unduly inconvenient. Drinks are always placed on the *right-hand* side of the person being served.

For persons *facing you* remember their right hand is opposite your left hand. It does not require to be worked out with a lot of gymnastics—mental, or callisthenic. Just pick up their drink in your LEFT hand and push it straight towards them—it will then be on THEIR right.

2
DRAUGHT BEER

The main lines of draught beer are Mild Ale and Bitter—in cask or canister.

MILD ALE

Mild Ale is referred to by customers as "Ale" or "Mild" and in some districts simply as "Beer"—meaning the same thing. It is known in the Trade as XX (Double X). The monks, who were great brewers, used the marks XX, XXX and XXXX to indicate quality and these marks still continue today.

For centuries Ale was the staple liquor brewed in this country and so important was it considered as part of the daily diet that an Assize was held to determine the price at which it was to be sold—based on the result of the harvest of the year. It was known as the Assize of Ale and Bread. Richard Whittington (1358-1423) as Lord Mayor of London presided at one such Assize and roundly berated the brewers for "ridinge into ye countrie" ahead of the bakers and buying up all the best crops.

The trend in recent years has been away from Ale or Mild and towards Bitter which many can now afford and consider much better value for money. In most public-houses mild ale is not even on sale and apart from a few elderly old-timers who still stick to their pints it is more often sold mixed in the following forms:

Light and Mild
Half-a-pint of Mild Ale and a bottle of Pale Ale served in a pint glass.

Brown and Mild (also called "Up and down" Brown)
Half-a-pint of Mild Ale and a bottle of Brown Ale served in a pint glass.

Stout and Mild

Half-a-pint of Mild Ale and a bottle of Stout served in a pint glass. (If more than one Stout is on sale it will be necessary to enquire which one is required.)

Guinness and Mild

Half-a-pint of Mild Ale and a bottle of Guinness served in a pint glass.

Mackeson and Mild (also known as "Mac" and Mild, or "Mackey" and Mild)

Half-a-pint of Mild Ale and a Mackeson served in a pint glass.

Mild and Bitter

Equal quantities of Mild Ale and Bitter served in a Government-stamped measure, either half-pint or pint as ordered. (If more than one Bitter is on sale it will be necessary to find out which one is required.)

Old and Mild

A strong old Ale, known as Burton (XXX or XXXX) and not so much in evidence these days, used to be a favourite when mixed with Mild. It is half-a-pint of Mild and an equal quantity of Burton (Old Ale) served in a Government-stamped measure, either half-pint, or pint as ordered. (Some breweries turn out a strong Burton as a winter drink—and very satisfying and warming it is in cold weather. Burton is the "old ale" called for in making Christmas puddings—if anyone *makes* Christmas puddings these days.)

(*The method of pouring mixed beers is given on pages* 28-9.)

Although Mild Ale (in London and in most parts of the South of England) is dark in colour[1] it should always pull up clear—crystal clear. If it is "murky" or "muddy" something is wrong and this should be pointed out immediately to the person in charge. *It should never be served to a customer*.

Draught beer which has remained in the service pipes during closing-time and has not been drawn off at the opening of the next session may be served to the first custo-

---

[1] MacMullans of Hertfordshire brew a pale mild ale and Greene King who provide for some areas a "country" mild which is also pale are among the exceptions.

mer. It will doubtless be cloudy and smell quite strong. A small quantity drawn off (a pint or so, according to the length of run to the cellar) will usually leave you with a beer fit to sell, but it is a reflection on the efficiency of the house if this has to be performed after the bar is open and in full view of the customers. (It is the cellarman's job to see that all draught beers are "pulled off" bright and ready for service at the commencement of each session.)

Mild Ale should always pull up with a nice creamy head and this should remain on the beer. If the head disappears rapidly, leaving the ale flat, it may be out of condition, or for one of the reasons explained on pages 14-15, which apply equally to draught as well as bottled beer.

BITTER

Brewers take immense trouble with their Bitter—ensuring that it reaches your cellar in clean, sparkling and prime condition, and *you* in turn are required to see that it reaches your customer likewise. It must leave you in a clean glass with a good head, not too full so that it slops over, not too much collar so that you have it returned for "topping-up". Brewers are well aware of the damage to their good name should anything be amiss with their beer, to the extent of pulling it out and replacing it with a fresh brew if there is the least doubt about it, so there is absolutely no excuse for serving anything short of the very best. Keep a watchful eye on *every* glass of beer you serve and assure yourself that everything is right.

You may have two or more qualities of Bitter to deal with; in a "Free" house there may be eight or ten!

NOTE: At this stage of your career, with much to study and remember, a long treatise on the brewing of beer would be out of place. It is therefore omitted as being superfluous to your needs. But when you have an opportunity of learning more of the Brewer's craft—take it![1]

Bitter is sold "mixed" in the following forms:

Light and Bitter

Half-a-pint of Bitter and a bottle of Pale Ale served in a pint glass.

[1] There are excellent and easily understood chapters on Brewing in *Innkeeping,* published for Brewing Publications by the publishers of this book.

Stout and Bitter (also rudely known as "Mother-in-Law")

Enquiry will have to be made here as to which Bitter and which Stout are required if more than two are on sale. Otherwise the service is the same as for Light and Bitter.

Guinness and Bitter

This is a proper *Black and Tan*—a name reminiscent of the Irish "troubles". However, care must be taken when a customer asks for a Black and Tan because in some places it means Guinness and *mild* or even *cider*. It is, of course, half-a-pint of draught and a bottle of Guinness served in a pint glass.

There is no end to the combinations of drinks with which customers punish themselves, but the foregoing is a list of those quite commonly met.

(The method of pouring and serving mixed beers is given on pages 41-2.)

KEG, or CANISTER BEER

Keg, or Canister Beer, which is by no means a "gimmick", has certain important features to recommend it.

It is delivered to your cellars in sealed canisters which cannot be tampered with. This is a guarantee that beer drawn from these canisters will be served in perfect condition.

A cylinder of CO_2 is connected through a reducing valve by a flexible pipe to a unit on the canister and this forces the beer up to the service point under pressure. Keg, or Canister Beers are served from special dispensers attached to the bar and are no more trouble to serve than the pouring of a glass of milk provided the pressure is correctly adjusted.

It is impossible to mention the hundreds of brewers marketing "Keg" or "Canister" beers up and down the country—all well known in their own particular areas.

BASS and WORTHINGTON

The reputation of these two great beers is jealously guarded by the brewers—aware that they are, perhaps, the "ultimate" in draught bitter. The sign "Bass from the Wood" is one of the greatest draws in the Trade. Casks are always referred to as "wood" and experienced bitter drinkers will persistently aver that beer drawn straight from the cask is superior to any drawn through pumps, or by any other method. This is an opinion constantly voiced and probably stems merely from the fact that they can *see* it being served.

Drawn through perfectly clean pumps, however, there seems to be no reason why the same beer in a cask in a good cellar should not be equal to it.

Worthington "E" has a great name and is preferred by some to Bass, but both these beers are of the same gravity and enjoy a tremendous following.

Bass and Worthington "E" are also racked and supplied as a Canister beer and this may even further enhance their popularity among Bitter—and Keg, or Canister Beer—fans.

PREPARING DRAUGHT BEER FOR SERVICE

There are five ways in which Draught Beer is despatched from the brewery and because this may have some bearing on the way it is served and its general appearance in the glass it is necessary to distinguish between:

Fined Beer
Unfined Beer
Chilled and Filtered Beer
Pressure Beer
Tank Beer

FINED BEER

A Fined beer is that which contains finings. Finings are a thickish, gluey substance looking rather like diluted paper-hanger's paste and are produced from isinglass. Isinglass is made from the swimming bladder of the sturgeon and other fish, and from this stems the name "fish-guts" by which finings are sometimes called by lower type cellarmen. Actually they have no offensive smell whatever—but they *do* deteriorate and *must* be used by the date stamped on the label of the container, or thrown away.

The object of fining beer is to leave it bright and sparkling. The finings, when mixed with the beer, gradually drop to the bottom of the cask and in doing so take with them yeast particles, hop leaves and pips and other substances in suspension in the beer. It follows, therefore, that while this operation is in progress the beer is not fit to serve and must be rested.

Just prior to leaving the brewery the finings are added to the beer, the journey of the dray and delivery to your cellar agitating them so that they are well mixed. The casks are then placed on Stillions (or Stillages) in the cellar and scotched up. (Stillions are the heavy wooden stands on which casks are rested and scotches are the wooden wedges

21

used for holding the casks steady on the stillion.)

After several hours the finings will have done their work, the beer "falls bright" and is ready for serving.

Once the finings have worked to the bottom the cask must on no account be moved (except by an experienced person and even then only very carefully) for if the finings are disturbed there is every possibility that they will rise and not settle again, leaving the beer thick, unpalatable and unsaleable.

Cellarmen, of course, are obliged to tilt the casks in order that all the available bright beer may be drawn off, but this is done very skilfully and with due care, using either scotches or one of the various kinds of tilt.

It will be obvious from the foregoing, therefore, that pulling on the pumps when a fined beer is low in the cask will result in a cloudy glass for the customer. More than this, it may mean considerable waste in clearing the pipes and pumps of the unsettled beer—and this is one of the reasons why you should look carefully at *every* glass of beer you serve.

UNFINED BEER

This is beer delivered by the brewery together with the necessary finings, fined on the premises as and when required. This is, however, a most uncommon practice nowadays.

This beer will keep and when about to be prepared has the shive (see Glossary p. 173) chopped out. After this a small quantity of beer is drawn off from the cask and mixed with the finings which are then poured back into the cask and vigorously stirred with a suitable stick (usually a clean broom-handle) so that they become well mixed in with the beer. A "fining-pan", usually enamelled and looking like a flat funnel, is inserted in the bung-hole which it is made to fit. The beer then "works" into the pan the excess yeast, etc. and when it is "quiet" a wooden plug is used to stop up the hole in the fining-pan which can then be removed from the cask without risk of spilling—and so disposed of. Several hours later the beer will have "dropped bright" taking the finings and all matter in suspension to the bottom of the cask.

The fining of beer is no slipshod operation and requires the services of an experienced cellarman. Probably one of his most important duties lies in judging the sale of beer so that the new cask is standing ready fined as soon as the old one "goes off". You cannot sell an unfined beer. You cannot fine a "working" beer.

Remember that "finings" are not part of the beer itself but merely a part of the preparation to render it fit for drinking.

Chilled and Filtered Beer

Most breweries now deliver their draught beer racked, fined and filtered. These beers have been freed from all sludge, hops, excess yeast, finings and other sediment. This type of beer will be clear to the very last drop and may be served immediately on arrival—even after being kicked round the cellar floor. This is a very definite advantage where there is a quick sale but this type of beer has no great keeping qualities unless protected from the atmosphere by a blanket of carbon dioxide admitted to the cask as beer is withdrawn.

Pressure Beer

Canister, or Keg beer is delivered in five or ten gallon containers or pins and firkin sizes. Each has a metal syphon with a valve at the top affixed to it. A unit to which the gas and main beer supply lines are attached is plugged into the top of the canister which is then ready for service—an operation which takes less than five seconds. Canister beer is ALWAYS sent out racked, filtered and ready for immediate use. In some cases no cylinder of gas is used and consequently there is no gas line—the beer having been "gassed" in the container before leaving the brewery. Thus there will only be a beer supply line.

Any difficulty experienced with these beers is usually connected with excessive pressure—or lack of it. Sometimes the reduction gauge does not operate properly. Six pounds per square inch is the usual pressure required at the bar, but up to sixteen pounds may be required to bring it from the cellar to the bar—where it is reduced to six pounds. It is better not to fiddle about with this apparatus, leaving any adjustment to the cellarman.

Tank Beer

This is beer delivered in bulk by road tanker and pumped into the cellar by pipe-line after the manner of petrol delivery to a filling-station.

23

After being filled a metal cap is screwed on and a gas line of CO_2 is connected. The tank now stands available for use as soon as it is required.

The changing over from an empty to a full tank is effected by releasing the gas in the empty tank to allow the supply line to be unscrewed, taking off the metal cap from the new tank and fitting in its place the beer supply line and turning on the gas. The method of changing over is simple in the extreme and is performed in a matter of seconds.

Another great feature of the pressure raising system is that the cellar can now look like a clinical laboratory.

Again, none of the beer is spilled on the floor as is usually the case with draught beer in cask when being prepared for service. Beer spilled on the cellar floor is the enemy of the conscientious cellarman and the bane of brewery Cellar Inspectors, for bacteria breed rapidly wherever it is found. As pressure tanks and casks with blanket pressure are entirely enclosed no bacteria can enter to attack the beer. With traditional cask beer air (always contaminated!) is of necessity attacking the beer through the small hole in the bung— drilled there for that very purpose, for without an inlet of air no beer could be drawn. So instead of an ugly stillion, a load of scotches, an array of taps, a kit of cellar tools, drip cans, wall tilts and other clutter, there is an air of spotless cleanliness in the modern cellar about as far removed from the old days as smoke signals are from Telstar.

DRAWING AND SERVING DRAUGHT BEER

Tilt the glass away from you. Bring it up to the nozzle of the pump, or tap, without allowing it to come in contact with it. Allow the beer to run down the further side of the tilted glass and as it fills gradually straighten it out to the upright position. This method of drawing applies to all draught beers in condition. If, however, you happen to be serving a beer which is out of condition, excessively cold or flat, it will be found advantageous to hold the glass well below the spout and, using pressure, endeavour to produce a "head" of sorts. On beer pumps the use of a "header", or "sparkler" may be found to help. This is an attachment fitted to the nozzle of the pump and has the effect of aerating the beer to give it appearance of a genuine "head".

Pulling up the beer too gently (from the pump) is a beginner's fault, as is pulling up too strongly and producing a glass of froth. A short pull to start with will show if the beer is likely to hold a "head", or have too much, and a little experience at the pump will enable you to adjust the strength of your pull accordingly.

NOTE: A properly adjusted beer pump draws half-a-pint with one straight pull. The flashy jerking of the pump handle backwards and forwards serves no useful purpose and may be dismissed as exhibitionism.

Beer drawn from the wood, i.e. from a tapped cask in the bar, will often pour out flat, or with a very thin head, and remain so, but this is in no way detrimental. Many experienced drinkers look twice at beer drawn from the wood which has a head and suspect that something is wrong with it. On the other hand those not used to drinking it will often complain that it is flat and also think something is wrong. Nothing is wrong. However, pressure may build up in warm weather and the beer may then draw in lively fashion and *will* show a head.

Watch the Overspill (see also page 27)

It is a sign of poor barmanship if you have your drip-cans full of beer. If the beer is exceptionally "high" due to hot weather, etc. overspill is sometimes unavoidable, as the excessive froth does not allow a full glass to be drawn and overspill becomes inevitable. However, it is too often a sign of slipshod service and can be cut to a minimum with a little care.

Don't keep pulling on the pump (or keep the tap running) if the beer is very frothy—just leave the glass for a second or two to allow the aeration to subside and THEN top it up. it is not uncommon to see barmen pull and pull and pull on the pump in an endeavour to draw a pint, whereas a short wait for the beer to find its level would be far quicker— and much less wasteful.

As previously explained you must always look at any beer before passing it over to the customer, draught beer certainly—bitter particularly—to ensure that it is in brilliant condition. Should it show any tendency towards "haze" or cloudiness *do not risk serving it*. See the person in charge and find out the reason. It is unfortunately true that some few publicans couldn't care tuppence about the appearance of their beer. In some houses the bitter looks like diluted cocoa and the customers (if any!) stand gazing glumly at it if they don't just feel in the mood for making a complaint, or realise it would be of little use. Such an establishment

COULDN'T CARE TUPPENCE

soon loses trade to competitors who consistently serve a brilliant pint.

If the beer is being drawn by pump from a barrel in the cellar it may be "low", i.e. on the bottom of the cask and be cloudy in consequence. Cellarmen do not usually sit waiting in the cellar for a cask to go "off". They may know it is low, but until somebody says something about it they would not know it had run too low to be served. But, regretfully, it must be observed that all too often a hazy glass of draught beer indicates that overspill has been worked back into the cask without being properly filtered.

It is a curious commentary on human nature—British human nature in particular—that a man will stand for almost anything rather than be caught (or think he has been caught) over his pint, especially on the vexed question of "collar". He demands an absolutely full pint "pressed down and running over", yet at the same time insists on a nice head. Very little thought is required to see that he can't have both (except amongst the Legal profession and the Weights and Measures Department who seem to think he can!). Usually there is some sort of grudging compromise —a slight head and less beer, but it should be known that proceedings have been taken recently by the Weights and Measures Department in a case of what they deemed excessive head. The Licensee was fined. Another appeal is pending at this time which may result in a sensible ruling from the Courts.

In connection with the above see also remarks on "Line" glasses (page 32).

WASTE AND OVERSPILL

Overspill

This, as the name implies, is the beer spilled over when drawing up for service.

In a bad class of house overspill is removed to the cellar, filtered, and worked back into the cask for re-service. The "guvnor's" attitude may be—"Well, it's good beer, it's my beer, it costs money and I don't see why I should lose over it." This is a business viewpoint and in the case of several pints of beer can indeed represent a serious loss when multiplied by fourteen sessions per week. However, it should be thrown away.

Waste

This is the bottom of bottles and the beer left in customers' glasses. Unfortunately there are still Licensees so lacking in thought for public hygiene that they filter and work back their "waste" and overspill into the casks for sale. This is a filthy, disgusting, and illegal practice, which cannot be too strongly condemned. It needs little imagination to think of the contamination suffered by such beer. In re-selling this beer the profit is depressingly small compared to the damage it may do to innocent people and to the good name of a respected Trade. Every Licensee has a moral obligation to ensure that this revolting practice is never allowed on his premises. The only beer which should be returned to cask is the "pull-up" at the beginning of a session or before cleaning the pipes.

General

The question of the disposal of genuine overspill is one of ethics. The scrupulous "guvnor" realising that the beer may have stood in the drip-can for several hours, contaminated by tobacco smoke, germs, staff coughs and sneezes will throw it down the drain; but this is where your assistance is required. See to it that the disposal of overspill does not arise in your bar by being very careful in drawing up.

Quite apart from hygienic considerations never try to "pass off" overspill (or waste) to a customer. If you are caught you may have reason to regret it. Obviously you are governed by the instructions issued by your employer, but if his attitude offends you you may have found one of the reasons for the popularity of bottled or keg beer in his particular bar.

NOTE: At the end of each session your drip-cans should be clear and you should then wash them out thoroughly in *hot water* and replace them *upside down*. This prevents dust from sweeping-up, etc. settling in them for not one in a thousand thinks to wash out their drip-cans at the *beginning* of a session.

POURING AND SERVING MIXED BEERS

The serving of mixed beers often gives rise to comment where one of the items is bottled. Strictly speaking one can *not* serve a "pint" of "Stout and Mild" or "Brown and Mild" etc. because, as previously explained, the Stout or Brown is not sold by measure and bottles vary in size and content.

The correct method to employ is to draw a **measured half-pint** of the draught beer, pour it over into a pint glass and hand the customer the bottle. Should the mixture fail to measure up to a full pint it is up to the "guvnor" whether he agrees to "top" it up at his own expense.

Some customers (and some "guvnors") insist on having the bottled beer in the glass first and then filling up with the draught beer until they have a full pint.

This is not correct

because the carbonation in the bottled beer causes it to froth up when pressure from the draught beer pump or tap is exerted. In some obstinate cases it happens that nearly a pint of draught must be pulled up to produce a pint glass of "mixed".

Two undesirable features of this method are:

(a) The draught beer, being under pressure, tends to

force the bottled beer to the top of the glass with the result that the overspill will be the dearer bottled beer and many customers remark on this in no uncertain terms.

(b) It is quite obvious that the overspill in the case of, say, "Black and Tan" (Guinness and Bitter) will certainly not be all bitter and this could very well spoil a couple of pints of otherwise good bitter already in the drip-can.

Mild and Bitter, "Mixed", "Half and Half"

There is no hard and fast rule as to which goes in the glass first, nor is there anything which says that the proportion of each *must* be fifty-fifty.

It was common practice at one time—when "Mild and Bitter" was more in vogue than it is today—for the bartender to pull up about two-thirds of a pint of mild ale and top it up with bitter; which may account for the fall from grace of this once very popular drink.

The fairest and safest way to serve "Mild and Bitter" is to *measure* half-a-pint of mild ale into a pint glass and then fill up with bitter IN FULL VIEW of the customer. The virtue of this method is in the fact that the customer can *see* he is not being "taken on". He can see the all-important bitter being served.

Draught Lager

Since the first edition of this book sales have increased so rapidly that in many bars they equal or even surpass those of any one draught beer. All draught lager will be supplied as "Pressure Beer" and the remarks earlier apply, bearing in mind that the pressure with lager is usually a little higher. Lager is always chilled, so remember always to check that the cooler is switched on before you open up.

3
GLASSES AND MEASURES FOR BEER

GLASSES

**Before you make any attempt to serve draught
beer make absolutely certain you are using the
correct glass.**

It has been explained previously in the manual that all
glasses must be clean and without a crack or chip, but of
utmost importance is the *size*.

**You have a definite responsibility in law to see
that all draught beer is sold in quantities of
one-third, one-half, or one pint.**

Pint and half-pint glasses are not just glasses—they are
Measures, Government-stamped by the Weights and Measures Department, and subject to inspection by their officers
on your premises to ensure the law is being strictly observed.

Normally you will find four basic types of beer glass
behind your bar, excluding odd shapes used for serving
lager, etc. They are:

The "Pony" or "Nip"

This is designed to take a "baby" beer in bottle, i.e. a
"Baby Ben" (Truman's) or a Bass No. 1 (Barley Wine) etc.

No legal responsibility attaches to the service of "baby"
size beers in *any size glass*, because bottled beer is sold by
the "bottle" and not (at time of writing, at least) by any
stated measure.

The "Twelve Ounce" glass

This is designed to take a small bottle of beer and leave
room to accommodate a nice "head". These glasses are not
Government-stamped. They are known as "bottle" glasses,

and in the case of stems, as "Worthington" glasses.

They are to be found in many shapes and designs, straight sided, tulip shape, stems (footed) etc. and they will all usually hold twelve fluid ounces (ten fluid ounces equals half-a-pint).

Again there is no liability in law governing service of bottled beer in the glasses, because bottled beer is *not* measured. You do *not* buy half-a-pint of Brown Ale—you buy a "small" bottle, and it is never sold to you as a half-pint. The public may think otherwise—but they are wrong.

Half-Pint Standard Imperial Measure

In *every* case where half-a-pint of draught beer is ordered it MUST be served in a Government-stamped measure—and there are no "if's" or "but's" about this! —whatever the customer says.

A trick often worked by the old hand is to pass a new barman, or barmaid, a pint glass about one-quarter full with the order to "fill this up with half-a-pint"—hoping, of course, that they *will* fill it up. Don't fall for it—it is illegal. The customer has ordered a half-pint and must be served with a measured half-pint, not some guesswork quantity. Even if he flourishes the glass in front of your face and repeats—"in here! in here!" just ignore him and draw it in a measured half-pint, then, if he says so, pour it into his pint glass.

If you do fill his glass up, apart from giving him a present of a lot more beer than he is likely to pay for, you are also giving what is known as the "long pull". The "long pull" is illegal. Many years ago, when competition was very fierce, publicans were known to give more than the exact measure when pulling beer in order to attract additional custom. This became known as the "Long Pull".

One-Pint Standard Imperial Measure

Once again these are measures, stamped as such, and *must* be used every time a pint is called for (but refer to paragraph on Mixed Beers, page 29, for an exception).

Pints may be glasses of various shapes, or mugs with handles, but in every case they must bear the Government stamp which indicates that they hold precisely one pint.

It is curious to note that in the West End of London and better-class establishments mugs with handles are more frequently used for draught beer than are glasses, whereas in heavy drinking areas the glass is more commonly favoured.

It sometimes happens that a customer has his own silver or pewter tankard behind the bar—for his own particular use. If it does not bear the Government stamp stating the capacity it is illegal to serve beer in it unless it has been measured first. The Weights and Measures Inspector may examine it on one of his visits and make some enquiries—so be clear on this point:

**Unless a tankard is Government-Stamped beer
must be measured into it.**

Line Glasses

These are oversize glasses marked to the level of a line etched on the side at the exact half-pint (or pint) mark and are stamped thus by the Weights and Measures Department. Although this type of glass was legalised in 1907 they have not been much in evidence until quite recently. In certain districts where they were re-introduced there was a great uproar from customers who, not noticing the "line", thought they were being caught over their draught beer as it did not reach the top of the glass. The are, of course, as anyone with an atom of sense can see, the obvious answer in serving a pint of liquid beer—and a "head" additional to it and not a part of it. Their use is becoming more acceptable with the adoption of the latest beer measuring meters, which deliver a measured half-pint.

MEASURES

CASKS, BOTTLES and STANDARD IMPERIAL

Intoxicating liquors sold in cask, or bottle, are only approximate as to quantity and there is no guarantee that they will carry the **exact** amount expected.

Beer in Cask

1 Hogshead	=	approx. 54 gallons, or 432 pints
1 Barrel	=	approx. 36 gallons, or 288 pints
1 Kilderkin (or Kil.)	=	approx. 18 gallons, or 144 pints
1 Firkin	=	approx. 9 gallons, or 72 pints
1 Pin	=	approx. $4\frac{1}{2}$ gallons, or 36 pints

Keg, or Canister Beer

Usually delivered in 5 or 10 gallon canisters, or pins and firkins.

DEPOSITS: The deposit a customer leaves on a bottle when beer is being bought to take away does *not* constitute a purchase. In other words the deposit *does not buy it*. It is, in actual fact, a sum of money deposited to ensure its safe custody and RETURN and at all times the bottle remains the property of the Bottlers.

Bottles, and especially syphons, cost very much more to produce than the deposits left on them.

NOTE: It is a serious offence to place anything of a poisonous, dangerous, or offensive nature in any bottle or container likely to be used again for consumable liquids.

4
SHANDY

This is the commonly used short name for "Shandygaff".
About the only thing in favour of serving a shandy lies
in the profit it shows! Usually called for on a hot day it is
considered to be a refreshing drink, but the time it takes to
serve (if others are waiting) takes off some of the gilt. A
smart barmaid will serve six Pale Ales, and thus satisfy six
customers, in the time it takes to dispense a pint of shandy.
Nevertheless it is a drink often requested, and it is your
job as a good bar-tender to try to satisfy all customers, so
grin and bear it, observing the following points.

Lemonade, or ginger beer, mixed with Bitter, or Mild,
makes a shandy. By the same reckoning if the overspill from
Lemonade, or ginger beer runs into your drip-can and is by
some mischance worked back into the draught beer then you
may have shandy in a big way, for it takes very little of
these two to contaminate quite a large quantity of draught
beer.

Pouring and Serving Shandy

It is important to draw the beer first and add the lemonade or ginger beer afterwards. The mineral water will froth
up alarmingly as soon as the draught beer is added, as it
liberates the gas, but by adding the *mineral to the beer* and
pouring it slowly down the side of the glass there is much
better control over the froth. It is almost impossible to avoid
overspill if you go to your beer tap *after* the mineral is in
the glass, but WITH THE BEER IN THE GLASS FIRST
you will be able to dispose of the overspill anywhere you
please, avoiding the drip-can.

Under the Excise laws it is a grave offence to mix different
kinds of beer together for re-sale. This comes under the
heading of *dilution,* and, of course, includes the mixing of
minerals with beer, whether this happens by accident or
design. The penalty may include the loss of the licence, so

every possible care must be taken to ensure that this never happens. (Where a customer has *ordered* "mild and bitter", or "shandy" the charge of dilution does not hold.)

As there are *four* ways of mixing shandy you *must* enquire if your customer requires Mild ale or Bitter with Lemonade or Ginger Beer. The commonest form of shandy is bitter and lemonade which is simply called "Bitter Shandy"—and this spotlights another reason for the *beer in first* directive. When a Bitter Shandy is completed it often looks very pale in colour and the customer may have reason to think he has been "caught" over the bitter—but if he actually sees it in the glass first then the question does not arise.

Another common call is for "Lemon and Dash". This is an almost full glass of lemonade with a "dash" of ale, or bitter, as required but poured out the other way round, of course, the ale (or bitter) in the glass first.

Ready Mixed Shandy

Often it will be found that people calling for a shandy are not much interested in its composition—they are seeking a rapid thirst-quencher. In these cases it is advantageous to be able to serve one of the brands of ready-made shandies direct from a bottle. *But you must satisfy yourself that the customer does not insist on the use of a draught beer.*

Ready-mixed shandy dispenses with the trouble of making-up, overspill—and waste of time. Particularly does it save a cross-examination as to the customer's requirements on its composition. Shandies are mainly called for during hot weather but are not always served as cold as the customer would wish. Ready-mixed shandy is usually dispensed from the "small" bottle similar to the beer bottle and this gives it the outstanding advantage over the normal shandy for the bottles can be kept on ice during hot weather.

As the alcoholic content of any drink containing not more than 2 per cent proof spirit is classified by H.M. Customs and Excise as a soft drink it would be safe to serve any young person of 14 or over with a ready-mixed shandy which fell into this category, but it would be most unwise to serve them with one made up in the bar as this might prove on analysis to be well over the prescribed alcoholic limit. (Refer to pages 142-3 on the subject of serving young people in bars.)

ENQUIRE THE PRICE TO CHARGE FOR SHANDIES...
BEFORE YOU START.

Size of Glass and Quantities

There is sometimes a doubt as to which size glass to use when mixing a shandy. Taking it logically: a *pint* shandy is served in a Government-stamped pint glass—therefore a small shandy should be served in a stamped half-pint glass. BUT! . . . for speed of service, and to avoid excessive over-spill, a twelve-ounce bottle glass is frequently used. If it is the practice in your bar to use bottle glasses for shandy make sure not to fill them to the brim, otherwise the profit mentioned at the beginning of this item will disappear!

Now a word about the quantity of beer to be used in making a shandy. Although this is "hit or miss" as a general guide it will be found to be about one-third of beer to two-thirds of mineral. Nothing can be laid down exactly as customers' requirements have to be taken into consideration, but if nothing is mentioned it will be found that this proportion is usually acceptable.

If your bar stocks two bitters ("best" and "ordinary"), you should also ask the "guvnor" which one, unless the customer specifies, you should use, and whether there is to be a difference in price. Usually the cheapest bitter is used, but best make sure.

It is important for you to make sure you know the price you are to charge for shandies, both bitter and ale, when you take over your bar.

No item in the whole Trade seems to be left so much to chance as the price, composition and size of a glass of shandy.

Prices are the cause of many disputes in the bar, so make certain you are *right* before you start.

5
CIDER AND PERRY

CIDER (or CYDER)

In Somerset, Herefordshire and Gloucestershire the spelling is Cider. In Devonshire and Norfolk the spelling is Cyder.

Cider is made from the expressed juice of the Cider apple suitably fermented, and is a strong alcoholic beverage produced in vast quantities in the counties mentioned above.

Records show that cider was made in this country in Saxon times. The wars with Napoleon curtailed the import of French wines to such an extent that cider production took an upward surge, and from this time the foundation of the flourishing cider industry was laid—by Napoleon! Another case of "good out of evil".

It may be recalled that in those days cider carried a Tax and beer did not. The taxation on cider grew so oppressive that cider drinking gave way to beer drinking. (At one time beer was considered on essential part of the diet of every schoolboy and was issued at meals to the boys of Eton—and other seats of learning.)

Cider is the wine of the apple. Both wine and cider are made in the same way, i.e. by pressing the fruit and allowing the juice to ferment naturally.

There are three classes of cider:

> Draught
> Bottled
> Champagne

and there are several varieties of the first two.

DRAUGHT CIDER

This is a cooling and most refreshing drink, deservedly popular and deserving of every possible care in serving. Like all good alcoholic drinks, however, its benefits can be abused and, because of its relative cheapness and high alcoholic content, it can occasionally present special problems in some areas; ask your Licensee, therefore, if there are any cider customers, or *would-be* customers, of whom you must be wary. In nine cases out of ten the answer will be no, but better to be sure than sorry!

Do, however, keep an eye on your unfamiliar draught cider customers, especially any who look bleary of eye or otherwise the worse for wear, because the drink itself is so pleasant that unwary youngsters (AND oldsters, too!) may not at first realise its alcoholic strength and suffer accordingly.

It is doubtful if you will have more than one variety of draught cider on your premises, unless you are working in cider-producing country, but in some cases two are kept. These are likely to be a "sweet" and a "rough".

Sweet Cider is produced by halting the fermentation artificially before the yeast has converted all the natural sugar into alcohol. Thus the finished product is controlled as regards sugar content from sweet to medium sweet, or medium dry. It will be a brilliant amber colour and as clear as crystal.

Dry Cider is produced by allowing the fermentation to complete itself—thus all the natural sugar in the apples is converted into alcohol. It is known as "rough" cider.

"Scrumpy" is a term used to describe a variety of "rough" cider—farmhouse cider. It is exceedingly dry and, not being filtered, is cloudy. It is, of course, alcoholically strong.

In olden days practically every farm had its own cider press and produced cider for the farmer and his workers. In the modern cider industry the word "scrumpy" is looked upon as rather derogatory and is not used by the big cider makers.

NOTE: Clean wooden taps (or stainless steel) are the *only* ones used for cider barrels. Cider should keep in good condition for about four weeks after tapping.

BOTTLED CIDER
This is bottled in the three standard beer sizes—and in Champagne quarts and pints. Bottled cider is filtered and will always pour out in brilliant condition. Unlike beer it does not have a "head" but it does have a fascinating

sparkle. The brand your house stocks obviously depends upon the size and class of your trade and the preference of your customers.

"MERRYDOWN"

The story of Merrydown is romantic and fanciful in the extreme. A business started on a shoestring (and a kitchen mincer) which has developed into a factory producing Cider and Fruit Wines in excess of 500,000 gallons annually could be nothing else.

Many are the stories current about the wonderful drinking times "the boys" have had on Merrydown—falling in the gutter, walking up the High Street on their knees, and punching policemen's helmets over their eyes. These "good times" have come about mainly through ignorance—ignorance of the alcoholic strength of this delightful apple wine —for the uninformed have treated it as an ordinary cider, which it is *not*!

Merrydown contains 23 per cent proof spirit and carries the full British Wine Duty of £1.30 per gallon.

The wine-like nature of this Vintage Cider has prompted the makers to re-name it *APPLE VINTAGE WINE* which is a sound and sensible idea for it should be drunk as a wine and not by the pint as cider. It is a potent drink to be treated with the utmost respect, and the advice given previously concerning customers thought to be drinking more than is good for them must be strictly adhered to.

It should be served, like wine, in wine glasses only.

PERRY

Perry, as such, is little heard of in the Public-house these days. It is the fermented juice of *pears* as opposed to apples in the case of cider. The Perry Pear is crushed and the resultant juice fermented in exactly the same way as the juice of the Cider Apple.

The most widely known Perry found nowadays is the much advertised product of Showering of Shepton Mallet— "Babycham". This Perry, so popular with the ladies, is improved by being served in a footed glass and adorned with a cherry (left on a cocktail stick, of course!). Two spots of Angostura Bitters (for details see page 109, Part II) will turn it a rosy pink and this tends to create interest amongst other customers.

"Babycham" is bottled in two varieties—sweet and dry. The sweet is the more popular and bears a blue label and a blue foil capsule.

NOTE: These products are NOT bottled in any other size but "baby".

6
SOFT DRINKS AND MIXERS

These are many and varied. They are subject to stringent regulations in respect of public health, preservatives in food, etc. and every invoice carries a declaration of purity and quality.

For the purpose of this manual Soft Drinks will be divided into groups and each group will be dealt with separately.

CARBONATED TABLE WATERS

These are the "fizzy" drinks on sale in sweet shops and cafés, non-alcoholic, aerated, sweetened and flavoured—and selling cheaply. They include such drinks as:

> Lemonade
> Orangeade (under various names)
> Grape Fruit—sparkling
> Cream Soda (sometimes called "American")
> Ginger Beer
> Cherryade

and many proprietary lines under various trade names, including some popular American drinks:

> Coca-Cola
> Pepsi-Cola
> Seven-Up, etc., etc.

Apart from the American drinks three of the others are commonly used in the Public-house Trade mixed with other items, and to cater for this they are supplied in "split" sizes (7 oz.). Frequent calls are made for "Port and Lemon"; Gin and Ginger Beer; Rum and "Coke" and details of serving these follow.

NOTE: In many of these drinks (including gin and tonic), ice will be called for. Don't, however, take it for granted that your customer wants it; ask "With ice, Sir?" If the answer is "Yes", *remember always to put the ice in the glass before you pour in the spirit*. No one seems to be able to prove why it's better that way but most people accept that it is—like pouring milk in the cup before the tea.

Pouring and Serving "Port and Lemon"

This is Port wine and LEMONADE—not lemon squash, not lemon juice, not a slice of lemon—but "fizzy" lemonade!

Pour your measure of Port wine into a glass large enough to take some Lemonade. A 6⅔ oz. wine and spirit glass should be large enough—not a half-pint—unless the customer particularly asks for it. If the Lemonade is stocked in "splits" hand the customer the Port and the bottle and allow the customer to do the mixing to his or her taste.

On no account start pouring the lemonade into the wine unless you are absolutely certain the customer has no objection.

If, as sometimes happens, your bar is only stocked with large bottles of lemonade then you should pour out the lemon in a separate glass (the pony size is usually acceptable) and hand both the glasses to the customer.

Pouring and Serving "Gin and Ginger"

Enquire first if your customer requires Ginger *Ale* or Ginger *Beer* (in the South it usually means Ginger Beer) with a piece of ice. Elsewhere it may mean Ginger Ale—so make certain. Serve the gin (a single unless a double measure has been ordered) in a glass large enough to accept whichever mineral is called for, *putting the ice in first*. Some customers may wish to use nearly all Ginger Beer and this is supplied in larger bottles than the small "mixers".

DO NOT START TO POUR IT. Just hand the glass with the Gin in it and the bottle of mineral to the customer for him to dispense to his own liking.

Pouring and Serving Rum and "Coke"

This is a measure of Rum and a small bottle of Coca-Cola. It presents no problems but as before—let the customer do it.

Do not start to mix it.

Pouring and Serving a "Snowball"

This is Advocaat and Lemonade—and in this case it is correct to serve it ready mixed in a 6⅔ oz. wine and spirit glass. The reason for serving it mixed is that it requires stirring (the Advocaat being thick) and this can better be done behind the bar where a spoon should be available.

Before you start to serve it you MUST enquire as to the quantity and price of the Advocaat, which varies tremendously from pub to pub. Pour the Advocaat in the glass and top up with Lemonade—then stir briskly to make a frothy drink—thus "Snowball".

"Snowballs" are now on the market in small bottles ready mixed. They need a good shake before opening.

Apart from Shandies (which have been dealt with elsewhere) carbonated soft drinks will also be called for by the glass as straight drinks.

NATURAL MINERAL WATER

This is natural spring water of varying medicinal properties according to the attributes of the springs from which they are drawn and bottled. The names of these waters crop up from time to time, but hardly any of them are served in the public-house, with one or two exceptions:

> Perrier Water (French)
> Vichy Water (French)
> Apollinaris (German)

"Have a baby Polly" is a well-known injunction and this natural mineral water goes very well with whisky. It is not so "gassy" as soda water. Vichy Water is imported in half-litre bottles and is sometimes called for—often to offset the dismal effects of a boisterous night out.

The following (for reference only) are a few of the better known Natural Mineral Waters from the Continent:

> Hunyadi-Janos
> Contrexeville
> Evian
> Vichy
> Apenta, etc.

and from this country:

> Malvern
> Buxton
> Springwell, etc.

In the main these waters are sold in high-class Off-licences and it is quite likely that your pub will not even stock them—if they have ever even heard of them!

FRUIT JUICES—5 oz. bottles

These are products which have sprung into great prominence since the War and they are justly popular. The principal varieties are:

> Tomato Juice
> Tomato Cocktail
> Pineapple Juice
> Orange Juice
> Grapefruit Juice

and various fruit juices under different names.

Tomato Juice, Tomato Cocktail

Some manufacturers produce a pure Tomato Juice and customers may ask for Worcester Sauce to be added. A bottle of Lea and Perrins should be available in your bar for this purpose. It must be shaken thoroughly before being added to the Tomato Juice and as it is inclined to settle on the top of the drink it should be stirred with a spoon before being handed to the customer.

Tomato *Cocktail* already contains Worcester Sauce as well as vinegar, peppers, lemon juice, salt, etc. and this should be pointed out to the customer should he ask for Worcester Sauce to be added. Tomato Cocktail does *not* need extra Worcester Sauce but some customers seem to think they are missing something if they don't have it. Do not be too liberal with the sauce—it is pretty hot stuff—especially if there happens to be some already in the drink. Lea and Perrins, which is the only one to use, have a shaker top on all sizes of their bottles and six good shakes should be enough for anyone—except, perhaps, a Burmese fire-eater!

Whilst on the subject of Tomato Juice (or Cocktail) mention must be made of the call for:

"Bloody Mary"

This is Tomato Juice, or Cocktail, with a measure of Vodka added. It is served to the customer ready mixed and THE VODKA GOES IN LAST. The Tomato Juice, being of a heavy consistency, prevents the spirit from mixing properly if it is unable to rise and the Vodka cannot rise with the Tomato Juice on top. A full bottle of Tomato is used, the Vodka is added, and the drink should then be well stirred with a spoon before being handed over the counter.

Another mixed drink which includes Tomato Juice (but to which no fancy name has been given) is chilled Tomato Juice and Sherry. This is usually consumed in the morning as a kind of pick-me-up—though much more likely to knock you down. The Sherry used is Dry or Medium Dry and NOT first quality. It is, incidentally, a favourite morning drink in South Africa.

Pineapple Juice

This is sometimes used for mixing with Gin, but in the main it is served without the addition of spirits, as are the pure orange juice and grapefruit juices supplied in baby size bottles. If called for, however, serve the Gin in a 6⅔ oz. spirit glass and hand the customer the bottle of Pineapple, or other juice. *Do not start to mix them.*

General Notes on Fruit Juices

NOTE (a): They are always improved by being served chilled.

NOTE (b): They must ALWAYS be shaken before being served. The contents are inclined to separate if left on the shelf undisturbed for any length of time. As they are NOT carbonated (fizzy) there is no risk of explosion on opening.

FRUIT SQUASHES AND CORDIALS

Orange Squash, Lemon Squash, Grapefruit Squash, Lemon Barley Water, Lime Juice Cordial, Blackcurrant Cordial, Peppermint Cordial, Ginger Cordial, are the most common.

The following notes on Squashes and Cordials should be carefully studied:

NOTE (a): Orange Squash, Lemon Squash and Grapefruit Squash, not being carbonated, are also known as *"Still"*. Should a customer ask for a Still Orange he will mean diluted Orange Squash—not the "fizzy" bottled variety— unless he means a bottle of Orange Fruit Juice, but this information must be given at the time the order is taken. Usually it will mean **diluted** Orange Squash, details of which are given under this heading later.

NOTE (b): *Alcoholic* Peppermint Cordial and *Alcoholic* Cloves Cordial must *not* be confused with any of the other Squashes or Cordials as they are very much more expensive.

As a consequence of this the *alcoholic* varieties are sold at very much higher prices and in carefully controlled measures.

Always examine your bottles to see if they are, in fact, *alcoholic* or *non alcoholic* before serving. Then find out exactly how much you are to charge in each case and the amount you are expected to give. Do not wait for this information until you are about to serve a customer—obtain it first.

NOTE (c): Give all your squashes a vigorous shake before serving unless they have been in constant use. (Make sure the stoppers are secure first, as with some makes these do not fit particularly well and your customers will not be too keen on being sprayed!) The sugar, which is used as a preservative in fruit juices and squashes, always settles to the bottom of the bottle and requires to be agitated.

NOTE (d): Make sure you are not using a metal pourer, or a pourer with metal attachments, in your squashes. The acid in the squash will turn them green with verdigris in very quick time and once this has happened they can never be restored to their original condition. This contamination is poisonous. Porcelain or plastic pourers are the *only* ones to use with any drink containing citric acid—the acid found in lemon, lime, etc.

NOTE (e): Wash your pourers regularly to save them becoming fouled. For preference leave them in water all night.

NOTE (f): Give all your squash and cordial bottles a good wipe with a damp cloth each session. They get very sticky and most uncomfortable to hold.

Cheap squashes are made with saccharin instead of cane sugar and this can be noticed by the absence of the thick syrup at the bottom of the bottle—even after they have been left standing for a long time.

Squashes—Served with Gin
GIN and LIME (lime juice cordial), GIN and ORANGE (orange squash), GIN and LEMON (lemon squash), GIN and "PEP" (peppermint cordial).

These are simple drinks but are often incorrectly and badly served. Generally customers do *not* want half-a-bottle of squash or cordial with their gin to drown it, nor will the "Guvnor" thank you for serving it in this manner. A splash is usually requisite—unless the customer specially asks for a more liberal measure. About the same quantity of squash, or cordial, as gin is a fair and acceptable amount. Never forget the golden rule:

With squash and cordial the gin goes in the glass last!

Lest it be thought that this is just a piece of nonsense and that as long as the customer receives a full measure of gin, first or last, he can have no complaint, it should be explained that certain squashes and cordials are strong enough to hide the flavour of the gin at the first sip. Peppermint, for instance, if slopped on top of a measure of gin may kill the flavour altogether and the customer may suspect that he has been "done" over the measure of *gin*. After all, your customers want to *taste* the gin—they're paying a price for it.

It is true that some customers will require more squash or cordial than others. Respect their wishes. Again, others don't trust you anyway and want to make sure that they get *all* the gin they are paying for. In these cases draw the gin, place it on the counter in front of them and take the squash or cordial bottle with it—but don't say—"before your very eyes . . . !" or anything like that!

WARNING: Here take note of a cunning little trick often played by tight-fisted customers trying (as always) to get away with something. Having ordered a round of drinks which included a Gin they will return to the counter and ask you to "put a drop of Orange in this". Make them pay for it. It's odds-on they are trying to get the Orange for nothing. The very same thing will happen with Lager and Lime—they expect the Lime for nothing.

"Still" Orange, Lemon, etc.

For service on its own, as a soft drink, about one-fifth part of squash to four-fifths of water (approximately) will be found adequate and allow fourteen to sixteen half-pint drinks out of a full bottle.

Real Thirst Quencher

There is on the market an unsweetened Lime Juice, but as it contains no sugar as a preservative the keeping qualities are not too good and for this reason it is not often found. If it *is* available a most refreshing drink may be made by mixing a liberal measure of this with a measure of Lemon

Squash, or Grapefruit Squash, adding plenty of ice and topping up with Soda Water, or Tonic Water, in a pint glass. It is an unbelievable thirst quencher.

Ginger Cordial

Nowadays this is used for little else than making up a cheaper "Whisky Mac". A measure of ginger cordial and a measure of whisky—WHISKY IN LAST! (See page 82 of Part II for further details.)

Blackcurrant Cordial

This is used for little else than making up "Rum and Black" (as it is popularly called). A measure of blackcurrant cordial and a measure of rum. RUM IN LAST!

At one time ginger cordial, blackcurrant cordial and raspberry cordial were popular served with hot water as a winter warmer, mainly for children, of course, but they never were public-house lines used in this form.

Peppermint Cordial (Non-alcoholic)

This is used in Gin and "Pep" (popular name) and Rum and "Pep". A measure of peppermint and the GIN (or RUM) IN LAST!

Your customer will usually state if he requires Peppermint of the alcoholic variety, but failing these instructions, you would be quite in order in serving the non-alcoholic kind, which is usual.

NOTE: It is often difficult to hear the call in a noisy bar especially if the customer speaks with an accent, or does not remove his cigarette. Many a time Rum and Pep has been served when the call was for rum and black. This sometimes represents a dead loss to the house, if no immediate sale can be effected of the drink drawn in error. Always repeat these orders clearly naming Blackcurrant in full, or Peppermint in full, so that there is no mistake on your part. It is a bad habit to call these items by the short name.

FRUIT CRUSHES

These are merely fruit juices already diluted to the correct drinking strength before they are bottled. They are "still".

GINGER BEER

There are two kinds of Ginger Beer—carbonated and brewed. Brewed Ginger Beer used to be sold in stone bottles because the force of the fermentation was often too strong for ordinary glass which used to burst frequently. The stone bottle had the effect of keeping the Ginger Beer nice and cool—especially in hot weather.

MIXERS

Certain soft drinks are known as "Mixers" because they are usually sold for mixing with other items. They are:

Dry Ginger Ale
Bitter Lemon
Tonic Water
Bitter Orange
Soda Water

—and they are usually available in the following sizes:

Babies (5 oz.)
Splits (7 oz.)

In some cases they are also available in half-pints. Soda Water, in many bars, is also dispensed from syphons on the bar counter and it is unusual for any charge to be made for what is known as a "splash".

Dry Ginger Ale

This is often referred to as just "Dry". Thus you will receive an order for a "Scotch and Dry" which means a measure of Scotch Whisky and a "baby" bottle of Dry Ginger Ale placed beside it.

You must never put any mixer in the spirit—never!

It is for the customer to adjust the "Mixer" to his, or her, own liking.

Customers busy chatting and not paying atention to giving the order sometimes make a mistake and ask for a Scotch and Ginger *Wine* instead of a Scotch and Ginger *Ale* —or the order may be misunderstood by you.

If it appears that a mistake may be made in the order ask the customer again and emphasise the Ginger *Wine* part of it. ("Sorry, sir, but you *do* want Ginger *Wine*—not Ginger *Ale*?") He may ask if you are deaf, but you will be hard put to it to separate the Scotch and the Ginger Wine once they are in the glass and the result may be a write-off, if he claims you have mistaken the order.

Bitter Lemon

This is very popular with gin and has taken the place of Tonic Water with many people. It is quite straightforward as regards the service—a measure of gin and a bottle of Bitter Lemon placed beside it—*not in it!*

It is *not* customary to serve a piece of lemon with Bitter Lemon—but it is, in this case, for the customer to ask. You should, however, enquire if ice is required.

Bitter Lemon is quite a popular drink on its own—without the addition of gin—and this has been known to cause a little consternation with certain orders. For instance —a customer may call for a gin and tonic and a Bitter Lemon, and will be served as requested. He may then say he didn't want a Bitter Lemon—only a "bit o' lemon". With

51

this in mind *always* refer to lemon as a *slice* of lemon, or a *piece* of lemon—and repeat this back to the customer to avoid any mistake.

Tonic Water

This is the favourite "mixer" with gin. It contains a very small quantity of quinine sulphate (about half a grain per pint) and it is this which gives it such a pleasant "bite" and tang. Once again it *must* be repeated that you *never* pour any Tonic in the customer's gin—just measure the spirit and leave the bottle beside it.

It is usual when serving Gin and Tonic to enquire if
        (a) a piece of lemon
        (b) a piece of ice
or both, are required, providing always that these two items are ready to hand. (Ice is often appreciated even in winter!) Serve the piece of lemon on a cocktail stick—and leave the stick in it! Serve the ice with tongs, or a spoon, *never* with the fingers.

It is preferable for *you* to ask the customer if these two items are required rather than for them to have to ask you—as though requesting a favour. Such little touches, small in themselves, do much to create a good impression and also indicate that you know your job!

Except on sultry evenings it is unusual for ice to be served with "Scotch and Dry" or "Rum and Coke" but here again it is for the customer to ask, or help himself from the counter ice-bowl.

In warm weather *all* "Mixers" should be kept chilled so you must make sure that you do not use up all your cold stock without having made replacements. Customers will scream their silly heads off if, having been served with, say, three ice-cold Tonics they are suddenly given a warm one.

It sometimes happens that a person will order a Guinness and a Tonic Water. In nine cases out of ten they intend to mix the two together as a "poor man's Black Velvet" and require the Tonic Water in the bottle. Here you should enquire if the Tonic is required poured, or just left in the bottle.

Bitter Orange

Not so popular as the other "mixers" but it is called for quite often.

Coca-Cola, "Seven Up"

The two are excellent "mixers". Coca-Cola goes well with Rum—"Seven-Up" with Gin or Whisky.

Soda Water

Supplied in all bottle sizes and in syphons it is much used with whisky and brandy.

A syphon should always be directed down the side of the glass and only very gentle pressure exerted on the lever. Should a syphon fail to work just give it a gentle shake.

Whenever a new syphon is taken into use always give a short squirt in the sink to release any excess gas. If you fail to do this the customer may find his whisky on the ceiling, or in his face, due to the initial pressure.

When handing a soda syphon to a customer pick it up in your *left* hand and place it directly in front of you. It will then be on the customer's *right* hand. Work it out. If you pick it up in *your* right hand you will have it pass it diagonally across him to reach *his* right hand. In doing this you run the risk of knocking over his drink.

NOTE (a): If, after removing the crown-cork of any bottle of "mixer" or Fruit Juice a dark tarry deposit is to be seen round the top of the neck—it is *old stock*. It means that your cellarman has been stacking boxes of fresh stock on top of the old until one of the old ones finally comes into use. As most "mixers" are handed to the customer in the bottle you may receive complaints so tell the person in charge. Age rarely affects the quality of the goods, but it looks bad.

NOTE (b): Ice is sometimes supplied to public-houses in blocks, or half-blocks, by ice merchants. This is used for cooling beers, lagers, etc. and especially canister beer where there is no refrigerated cellar. Lager firms often supply ice tubs in which bottles are placed with ice packed round them as display items on the bar counter.

This is *commercial ice* and should *not* be used in drinks—*ever!* There is no guarantee about the state of the water from which it is produced—or what it has suffered on the journey in the ice cart. *It is not* intended for use in drinks.

NOTE (c): If you remove ice cubes from trays in the refrigerator *always refill the trays with water*. Nothing is more devastating to morale than to go to the 'fridge on a

'INITIAL PRESSURE'

really hot day to find the ice cubes gone—and the trays empty. If your customers "chew you up" about it they may well be justified.

NOTE (d): To prevent ice cubes from forming into a solid mass in your ice bowl give them a short squirt from the soda water syphon.

NOTE (e): To keep slices of lemon from going hard and dry do not cut too much at a time, but keep just a sufficient quantity in shallow water.

NOTE (f): It is not always realised that cocktail cherries, although given away, are not cheap. Don't treat them like pea-nuts. Although they may look tempting you will receive your meals at the appointed time.

PART II
WINES, SPIRITS AND VERMOUTHS

7
WINES

This is not intended as a fully comprehensive treatise on wines which would cover acres of paper and be of little benefit to you behind the bar. Unless you are called upon to serve in an "Off-licence" or a restaurant dispense, your range of interest will be limited to those wines sold "by the glass" in a normal kind of public-house. Nevertheless the subject is so vast that it cannot be dismissed in a few words, and the information given herein will always be found useful to those employed in the Trade.

Wine is a *living* thing. Treated by the uninformed and inexperienced as just another drink, they are often at a loss to understand why, for instance, it should go cloudy, throw a heavy sediment, or turn sour.

Wine breathes; it is subject to vagaries of temperature, it can catch cold, it can fall sick, it can wither and die!

The average public-house will probably serve no more than a dozen kinds of wine "by the glass"—the principal ones will be the Sherries, of which there are so many varieties. There will be two or three kinds of Port and possibly a South African, Australian or even British product.

However, much depends on the class of house and the type of trade for which it caters. You may have one customer who regularly drinks Marsala, or another who will call equally regularly for Tarragona—an event peculiar to the particular house. On the other hand, one could go for ten years without serving either of these by the glass.

Wine is divided into three main classes:
Fortified Wine
Table Wine
Sparkling Wine

FORTIFIED WINE

This category embraces Sherry, Port, Madeira and Marsala. The process of fortifying wine brings the normal fermentation to a halt—and for a special purpose—to prevent it becoming too dry, too harsh and too alcoholic.

This obviously requires some explanation, so the process of fermentation will be followed in very simple terms.

Grapes specially grown for their particular characteristics are collected in from the vineyards and pressed. When they are gathered they have on them a "bloom" such as is seen on hothouse grapes—a kind of white dust. This "bloom" is the yeast. When the grapes are pressed the grape-sugar stored within the grape becomes exposed and is attacked by the yeast.

Take a bunch of hothouse grapes with the "bloom" on them. Whilst they remain intact the "bloom" can still be seen. Remove one grape and within a short time the "bloom" begins to disappear starting with those grapes nearest to where the grape was pulled off, the stalk of which will still have some flesh adhering to it. This is the yeast moving in to attack the exposed sugar—minute though the quantity may be. Tear off two or three grapes and you will soon have no "bloom" left anywhere on the whole bunch.

Chemical changes take place as soon as the grape yeast sets about the grape-sugar and the result is—alcohol! The yeast will continue to attack the sugar until every particle of it has been converted into alcohol. To allow the fermentation to complete itself naturally would result in a coarse, unpalatable, dry wine, without any pretension to sweetness.

So far everything in the process has been done by nature —the sugar in the grapes has been turned into alcohol by the action of the yeast carried on the grape skins, unaided by any external influence.

Now comes Man, who, anxious to retain some of the sugar in the wine to give it sweetness, discovers that by adding a specially distilled grape spirit to the liquid the fermentation ceases—leaving some of the sugar unconverted. Thus he has what he requires—wine containing enough sugar to make it palatable, with the spirit he has added making up the alcoholic strength it lost through having the natural fermentation halted in mid-stream.

The process thus described is known as "fortifying the wine". The strength of fortified wines varies between 17 and 21 degrees alcohol (by volume) and the two most important fortified wines are Sherry and Port.

THE SHERRIES

It is safe to say that no wine causes so much confusion as Sherry, among amateur barmen or barmaids, who rarely have any idea what they are serving—or what it is like to drink. For this reason it is one wine worth studying, as every often a customer will be looking for advice and will expect it to be more than mere guesswork.

Sherry comes from Spain and nowhere else! It is produced from white grapes picked with great care when they are at the peak of perfection. After being laid out on straw mats in the sun they are crushed, and the juice or "mosto" is run off into casks and allowed to ferment naturally until the November following the vintage. Each cask is then tasted and classified as to quality and type. Colouring is by the addition of a wine known as "Vino de Color" and sweetening by another wine made from very sweet grapes. Being made from white grapes Sherry is therefore a *"White Wine"*, a point worth remembering.

Sherry is a *blend* of wines—of various years and styles. Blending is designed to ensure uniformity year after year, otherwise shipping would be a most precarious business— the buyer being uncertain as to the type of wine he would be presenting to his customers, and the shippers not knowing if they could continue the supply of any particular wine; so Sherry is blended—and it follows, therefore, that there are no Vintage Years as in the case of Port. However, it is possible that you may find a "dated" Sherry prefixed by the word "SOLERA".

The bulk to which the new wine is added is known as "SOLERA". Solera is not a wine, or a kind of wine. It is merely the foundation on which the blend is built. The word comes from the Spanish "suelo"—meaning ground or floor. Thus a label bearing the legend "SOLERA 1868" means that the *original* vat was laid down in that year.

Year by year a quantity (up to about 20 per cent) has been drawn off and replaced by a like amount of *newer* wine as near to the bulk in style and quality as possible.

NOTE: In Spanish the double "l" is always pronounced like the "ll" in million. Therefore the correct pronounciation of Montilla is Mont*eel*-ya and of Amontillado—Amontil-*yah*-do. The stress is on the last syllable but one, thus, Olor-*o*-so; Glori-*o*-so.

The variation in style of Sherry is so great that it is never safe to accept an order for—"a glass of Sherry!"—unless you know beforehand which kind is required. This sort of thing is liable to happen when one of a party is including the Sherry in a round of drinks. A customer being served a Pale Dry variety might describe it as "poison" if used to drinking Rich Old Brown.

The pale Sherries are dry, the golden ones vary from medium dry to full rich and the brown are rich dark and sweet.

Dry means unsweetened, but not, of course, sour. Sweet is self-explanatory.

Now, it may happen that you will come across a Sherry with just a name on the label and no description. The name will very likely be in Spanish and leave you without a clue as to whether the wine is sweet, dry or medium. Sometimes even a price-list will not disclose the information and this is one of the shortcomings of the Trade generally. The following list will help you to identify the type of wine in such cases.

Fino (Pronounced Fee-no)

The Finos are pale and dry. They are often blended with other types of wine to produce dry clean Sherry, but they may sometimes be sold as "Fino".

Golden

These range from light to full rich. Therefore a "Pale Golden" would be dry, yet not as dry as a Fino and a Full Golden would be darker and sweeter.

Vino de Pasto (In Spanish—"Wine of the Repast")
As the name implies, this is a dinner wine, and, as such, is pale and fairly dry.

Amontillado (pronunciation on page 81)
This is a blend of Fino wines and has a "nutty" flavour. It is usually classed as being medium, i.e. not too sweet, nor too dry.

Amoroso
This is a rich golden wine.

Oloroso
Dark and full flavoured. Both Amoroso and Oloroso are full-bodied and "generous" wines.

Cream
This is a blend of fine Oloroso wines, usually full and rich and golden in colour.

East India
At the beginning of the nineteenth century, it was the practice to despatch sherry in casks on voyages to India, or even round the world. The rolling of the old sailing ships improved the wine and consequently it was much sought after—at a high price. The words "East India" do not denote any particular *style* of sherry.

Tio Pepe (Pronounced *Tee*-oh Peppy)
This is a blend of selected Finos of great age skilfully married to produce what is generally accepted to be one of the very finest dry sherries, Tio Pepe is the largest selling dry sherry both in Britain and in Spain.

Dry Fly
A fine medium dry sherry suitable for all occasions and all times of the day. Light and delicate with a fine golden colour. Shipped exclusively by Findlaters.

Double Century
The firm of Pedro Domecq have been shipping fine sherries to this country for over two hundred years and this fact gives the name to a fine wine on sale in almost every public-house—*Double Century*. Its popularity is justly deserved as it is an "all purpose" sherry.

Celebration Cream

Another sherry from the vineyards of Pedro Domecq, it is a blend of fine Oloroso wines of great age and distinction, resulting in a pale, full-bodied, mellow wine, produced to meet the every-growing demand for "Cream Sherry". It may be recommended with every confidence.

Bristol Cream, Bristol Milk, Bristol Dry (Harvey's)

Wherever wines are sold, Harvey's Bristol Cream will certainly be called for. This does not mean only in superior City bars, but in ordinary public-houses—and "back street" ones as well!

Harvey's of Bristol have built themselves a reputation which even the working man respects if he appreciates sherry at all. This may come as a surprise to many people new to the Trade, but it is an impressive tribute to the discrimination of customers who nowadays can afford the best —and insist on having it. Doubtless, Bristol Cream will be the most expensive Sherry on offer in your bar.

The bookmaker who orders Champagne (of any kind) to celebrate the losses of his unfortunate clients and to make a big impression on the assembled company, has nothing in common with the man in the cloth cap who orders a glass of Harvey's Bristol Cream in your Public Bar. The Sherry drinker consumes his wine for his own satisfaction—not with any ulterior motive. He demands value for money, which probably accounts for the increasing popularity of Harvey's Bristol Cream in all bars.

Both Bristol Cream and Bristol Milk are full-bodied Sherries which are blended on the solera system in the Harvey bodegas from Olorosos of great age. Bristol Cream was developed by Harvey's in the nineteenth century as a dessert Sherry blended with even older Olorosos than that used for Bristol Milk.

Bristol Dry, a medium dry blend of old Fino, is of particular interest because it was the wine which introduced Britain to chilled Sherry. As stated earlier in this manual, Sherry is a white wine, and the taste of the paler kinds, such as Bristol Dry, is improved by slight chilling before serving. In Spain, this has long been the custom, and, thanks to Harvey's, the habit is rapidly spreading in Britain. But note that the sherry should be chilled only—not frozen.

NOTE: One point here must be made in connection with Harvey's wines which has nothing to do with the wine itself. The bottles are stoppered with a very handsome cork—a round ball, covered in black plastic material. The easiest way to open a new bottle is to cut the plastic covering just underneath the knob. However, the cork, when replaced, can give the impression that the bottle has not been opened, so perfect is the fit, and it has been known for a stocktaker to write a bottle into stock as full and unopened when, in fact, it was found to be nearly empty.

If you are present at stocktaking point this out to your stocktaker, remembering that if you are £1.50 over this time you must be the same amount short on your next stock —and the chances are you won't hear much about being over!

NOTE: Because two sherries are labelled "Pale Dry" it does not follow that they will be the same either in colour or style and therefore it would be very unwise to substitute one "Pale Dry" for another. A customer may have been drinking Jones's Pale Dry—and because it runs out, the bartender may be tempted to serve Smith's Pale Dry in substitution. *It must not be done.* Sherry drinkers usually have a very perceptive palate, and may well remark on the difference, causing embarrassment. In any case this comes under the heading "Passing Off" and is an offence over which there have been long and expensive lawsuits. It does not matter if one is as good as, or even better than the other, it is still "Passing Off"—so the customer *must* be clearly asked if substitution is acceptable.

The sherries of Williams and Humbert deserve special mention in this manual as several of them are certain to be found on sale in your bar and each one carries the reputation of a firm shipping fine wines since 1877.

If your house enjoys a good sherry trade then you are quite likely to find the following also on display:

Pando: A blend of the very finest old Finos combining delicacy and charm resulting in a very dry sherry most suitable as an aperitif.

Carlito: A favourite Amontillado matured and blended to give the much desired dry finish.

Dry Sack: This is an original Sack wine (see page 66). It is not as dry as a dry sherry—but is a perfect example of a dry sack wine—and very popular.

Canasta Cream: Fine Olorosos of great age are blended to produce this rich cream sherry—a wine which may be confidently recommended when the call is for something not quite as rich as the brown variety.

Walnut Brown: This very old brown sherry will always be found acceptable when a sweet blend is required—the name itself being sufficient guarantee of satisfaction.

We must not leave the subject of Sherry without reference to the fine range shipped by Hijos de Jimenez Varela of Puerto de Santa Maria. The name "Varela" will come frequently before you on the shelves of thousands of Licensed Premises throughout this country and also in Scotland and Ireland. Wines of the "Varela" range stand solely on the merit of the quality which results from judicious blending and decades of maturing in straight soleras. Respect for quality is a predominant feature of the wine drinker and any Sherry which enjoys a long and continuing popularity does so by virtue of one thing only—outstanding quality.

The "Varela" range includes:

Manzanilla: A dry wine from Sanlucar, very pale in colour and having a slightly bitter after-taste. For the discriminating palate this is an outstanding aperitif.

Fino: A superior delicate dry wine of outstanding quality. It possesses great character and has developed a smoothness and delightful bouquet from many years in bodegas.

Dry: A delicate wine for the appreciative drinker. It has a delightful bouquet and is difficult to better as an aperitif.

Amontillado: A very old wine of outstanding character. Superbly balanced and with good colour.

Medium: This rare wine is produced after years of ageing in a Royal Bodega at Puerto de Santa Maria. Neither too sweet, nor yet too dry, it possesses perfect balance and can be drunk at any time, anywhere.

Oloroso: A full rounded wine with a slightly dry finish, fine bouquet and good body. An eminently satisfying wine.

Cream: A rich luscious wine with a very mellow character. A very old wine of very high quality and exceptional smoothness.

Brown: A velvety, rich dessert wine of beautiful colour. Superior quality with good age and character.

Bearing in mind that the Sherry trade is a "nice" trade as far as the public-house is concerned and the people who drink it besides being appreciative are usually discriminating it is important to remember that those asking for "Walnut Brown", for instance, are not going to be satisfied with anything else—and it is a grave mistake to try.

NOTE: When arranging Sherries on your shelves *always* set them out in sequence—from the very dry, through the range, to the brown. You will find this a great help when selecting one to suit the palate of a particular customer. Don't run out of stock—and don't forget to replace the corks each time they are removed—not tomorrow!

The only other words you will find on the label of a bottle of Sherry (apart from the Shipper's name) will be "Jerez de la Frontera"—"Jerez" or "Xeres".

This is the town which is the centre of the Sherry trade. Though formerly Sherry was produced in a limited area around Andalucia, in the province of Cadiz, it is now extended to cover nearly all the vineyards in Southern Spain.

It is illegal to offer any Sherry for sale which has been made outside Spain without the name of the country of origin preceding it—thus "'Cyprus" Sherry.

It is from Jerez that the wine, and the name of the wine is obtained. In early days, all wines were sweet and many came from the island of Crete (then in the power of Venice) and also from most of the Aegean Islands and Cyprus. They were known as Malmsey. Trouble started early in the fifteenth century when the Venetians decided to increase the price of wine and refused to accept English woollens in exchange.

Henry VII took action by building a larger mercantile marine fleet. Our ships went to the Mediterranean to sell woollens at better prices and as a return cargo brought Malmsey back to England from places other than Crete. The Venetians became alarmed and levied taxes on all wines not carried in their own galleys, to prevent the wines being carted elsewhere and shipped in English bottoms. This impost was resented by the English who later placed an import duty on Malmseys brought into the country by foreigners as a temporary tax in an attempt to win the mercantile struggle for our ships.

Negotiations followed lasting many years which aimed at a repeal of the taxes by both sides. The Venetians finally capitulated and repealed their tax on the understanding that the English would do the same. But when the English noted that the position of Venice in the Adriatic was not too strong, they reduced the import tax but never remitted it as they had promised. Perfidious England!

A few years after Henry the Eighth came to the throne an offer was made by a Spanish overlord to English merchants, granting them various privileges as an inducement to buy the wines of Jerez and surrounding districts. These wines were natural wines and were very much drier than any previously shipped from Spain, or the Malmseys. Accordingly they were called "Seck" wines, from the Spanish "seco"—meaning dry, and later were known as "Sack".

As the popularity of Sack increased in this country many other places started to imitate it and "Sack" was shipped here from the Canaries and elsewhere. It therefore became necessary to distinguish between the Sack of Jerez and the others, so it was called "Jerez—Sack", later Sherris—Sack, and finally the word Sack having been dropped, Sherris became Sherry.

Sherry remained a popular favourite for three hundred years, even surviving the war of Elizabeth I with Spain— a few nice captures at sea helping!

Now, having come this far, it should not be difficult to translate a label printed in Spanish, thus:

La Querida = La Querida (the brand name)
Fino = Fino (a pale dry wine)
Hermanos Quintana = shipped by Quintana Brothers
Jerez = a genuine sherry from Spain

NOTE: Sherry being made from white grapes is a white wine and, in common with all white wines, is improved by being served slightly chilled, but not cold. This applies only to the *pale dry* Sherries.

Ice should never be placed in wine—any wine!

The following names represent a guarantee of quality: Domecq; Sandeman; Harvey; Duff Gordon; Gonzalez Byass; MacKenzie; Misa; Williams & Humbert; Garvey; Findlater.

Important Note
Replace the stoppers and corks on all bottles, especially wines, at the time of serving.

PORT WINE
Port wine comes from Portugal—and nowhere else!

As has been explained in the case of Sherry, woollens played a large part in the wine trade of former years. In the year 1703, the British Ambassador in Lisbon, Paul Methuen, negotiated a treaty of commerce between Great Britain and Portugal wherein advantageous terms were arranged for the sale of Port wine, and the purchase of English woollens. This is known as the Methuen Treaty and is a model of its kind in commerce. Under this Treaty, Port wine was admitted to this country carrying a duty of £7 per ton whilst the duty on French wines was £55 per ton.

A further Treaty, made in 1916, restricts the name "Port" by law and no wine not shipped through the bar of Oporto is allowed to be called Port—which in these days of blatant and plausible imitation is a good thing. The control is very strict on both sides.

For some reason quite unexplained, Port wine may be enjoyed in this country better than anywhere else in the world—including even Portugal itself where it is produced. It may be something to do with the climate, the damp air, or some other factor, but the fact remains that it is not the same in New York, Bombay or Paris.,

Port is a fortified wine like Sherry. It is made from black grapes which are fermented with the skins, the dye from which produces the ruby colour.

White Port is made from white grapes—but is *not* classified as a White Wine. Every white wine is improved by being served chilled; White Port is the one exception, and should not be served other than at room temperature, in the same way as the red ports. (See *Porto Branco* on page 95 for an important exception.)

For centuries the grapes have been trodden in huge troughs which has been found to be the best method, because the human foot does not crush the pips or the stalks and release the acid into the grape juice.

But along comes Progress, Time and Motion Study and all the hoo-ha of modern times and a little investigation reveals that this traditional method is wasteful of time and endeavour because the machine can do the job as well if not better, quicker and at less cost.

A centrifuge (big brother of a spin dryer) has been found effective.

So the grapes, instead of being crushed by bare feet, are now "spun", broken in such a way as to break the skins, extract the pulp but leave the pips and stalks intact.

Although "treading" continues, difficulties in recruiting the necessary labour and increasing wages may mean the eventual abandonment of this picturesque part of the vintaging of Port wine.

The juice procured by either method is allowed to ferment, but before all the grape sugar has been converted into alcohol, a quantity of mature Brandy is added and the fermentation ceases. Thus the wine contains a proportion of the natural sweetness of the grapes as well as the strength from the added alcohol.

Port is usually sold at about 35 per cent proof, and is therefore approximately half as strong alcoholically as Whisky or Gin at 70 per cent proof. Reckoned another way—a customer diluting whisky fifty-fifty with water would be consuming about the same quality of alcohol as is contained in an equal quantity of Port wine.

(For a full explanation of Proof Spirit see pages 114-17.)

There are six categories of Port, and they are described below—but only four of them are likely to concern you.

Vintage Port and Crusted Port are practically unheard of in the average public-house, but sometimes a bottle or two are to be found in the dark corner of a cellar, and it may be your misfortune to disturb and perhaps ruin them. For this reason a few words about these wines will not be wasted.

Vintage Port

A precious jewel to be handled with the utmost care. It is very unlikely that you will be called upon to serve, or handle this wine which can be completely destroyed by anything but the most delicate handling.

Vintage Port is the wine of a Vintage Year. Many factors combine to produce a Vintage Year—a Spring free from frost, which would damage the young vines, a hot Summer with fine rain to swell out the grapes, and a dry spell at the harvest gathering, so that none of the natural grape yeast on the skins is washed away.

These are only three items concerning the husbandry of the grape, but when every condition of climate has combined during any one year to produce a wine of exceptional character, then the Shippers may decide to call it a "Vintage Year". Here it is interesting to note that not *all* wine produced in a Vintage Year is good. Similarly some very good wine may be made in a non-vintage year.

The wine of a Vintage Year is kept in casks for about two years, and is then shipped here for bottling. After bottling it is laid down, a whitewash brush is run along the bottles, and from then on it commences to improve year by year in the bottle. During this time it throws a firm crust on the underside of the bottle which must not on any account be shaken or disturbed. Owing to this crust the wine must be decanted before being consumed, and this operation requires great care, and a very steady hand.

Vintage Port is sold under the name of the Shipper, and

the year of the vintage: thus, Croft (the Shippers) 1950 (the year of Vintage). Old Vintage Port is often *not* labelled —the corks, however, are branded with the name of the shipper and the date. The cork is the guarantee, and should always be retained for inspection by the purchaser. If being served at table, the cork is often tied to the neck of the decanter—as proof!

The purpose of the whitewash mark on Vintage (and Crusted) Port is to ensure that the bottle is replaced in exactly the same horizontal position after having been moved. The whitewash mark is, of course, always uppermost.

The Vintage Years this century: 1904; 1908; 1912; 1917; 1920; 1922; 1924; 1927; 1931; 1934; 1935 (1942, 1945); 1947; 1948; 1950; 1955; 1958; 1960; 1963; 1966; 1967; 1970. The '42s and '45s were bottled in Portugal—due to war-time shipping difficulties.

Crusted Port

These are high quality wines made in those years not quite good enough to be declared vintage years. They are treated in the same manner as vintage wine. They throw a crust in the same way as their exalted brothers, and they require to be decanted with the same care. They may be the wine of the year, or a blend of several years of matching quality.

Crusted Ports have the same whitewash mark on the bottle as Vintage Port. The corks may carry the name of the Shipper—but no date. They are not labelled.

NOTE: It is impossible with both Vintage and Crusted Port to decant the whole bottle. The crust and loose sediment may account for as much as 25 per cent of the bulk. Thus, a Vintage Port purchased for say £3 could work out at the equivalent of £4 for a full bottle! Usually, about seven-eighths of a bottle is drinkable.

(Decanting these wines is hardly likely to concern you and is therefore omitted from this manual.)

Vintage Character Port

These are fine old wines blended from the harvests of different years. They are racked from the lees before bottling. (Racking is the removal of the liquor from the dregs or sediment.)

These wines are sold labelled, and are bright in bottle—but if kept too long (as will all wines) they may throw a sediment.

Tawny Port

This is a blend of wine made from the grapes of several years. It is left in the Shippers' lodges at Oporto for many years to mature in the wood. This causes the wine to lose some of its colour—which is absorbed into the staves of the casks. It does not improve in bottle and is therefore shipped when ready for drinking, and is not "laid down". If kept too long in bottle it may lose its brilliancy and throw a sediment.

Ruby Port

This is sound commercial wine of a kind usually sold in the Public-house by the glass. It is kept in casks for some years during which time it loses much of its "fire"—and some of its colour—but nothing like so much as the Tawny.

It may be the wine of one year, but is more often the blend of different vintages designed to maintain continuity of colour and quality. It could be said to stand half-way between the Vintage and Crusted Ports, which are bottled early and thus retain their colour, and the Tawny Ports which lose colour through being kept many years in the wood.

White Port

Produced in the same way as Ruby Port, but from white grapes instead of black. The best known and most widely advertised White Port is "Clubland White" by Parkington, and is one which may be confidently recommended. Generally speaking, White Port is pleasant, and frequently sweet. Which brings us to:

Porto Branco

Sandeman's introduced an entirely new departure in the serving of White Port—as an aperitif! This firm has introduced a light White Port with only a suggestion of sweetness called *Porto Branco* which they recommend should be served chilled, or on ice.

As chilling is known to bring out the delicacy of white wines—and as White Port *is* a white wine (previously explained)—there is no reason why it should not be served off

71

the ice. In fact, given a fairly dry White Port, it seems to be the obvious way to drink it. The reason for serving it at "room temperature" has doubtless stemmed from its general sweetness.

Sandeman's also suggest it should be tried as a long drink, adding a sprig of mint, a slice of lemon, and filling up with soda water, or tonic.

Notes on Port

With both Ports and Sherries watch for a cloudy sediment at the bottom of the bottle. Do not attempt to serve the last glassful straight from the bottle—pour it into another glass first and examine it for colour. If it looks thick, cloudy and turgid do not serve it. Never pour the bottom of one bottle over into a new bottle; unless the wine is absolutely brilliant you could ruin the new bottle.

Although Sherry, being a white wine, does not suffer quite so much from a chilly atmosphere, nevertheless, it does not appreciate constant changes of temperature, and the brown ones especially may incline to cloudiness—so they should not be left near a hot-plate, or heater. Port, on the other hand, may go cloudy if left too long in the cold. Unless it has gone seriously "sick" its recovery may often be brought about by keeping it in a warm, even temperature for a day or two.

Red Port should always be served at about the same temperature as that of the room, or bar, in which it will eventually be consumed. In other words, it should not be brought up from a stone-cold cellar and opened for immediate service. On no account should it be stood in front of a fire, or plunged into hot water, in order to achieve the desired rise in temperature.

The following are some of the first-class shippers of Port Wine:

Dow; Cockburn; Fonseca; Croft; Sandeman; Warre; Rebello Valente; Taylor; Gonzalez Byass; Delaforce; Graham; Hunt Roope; Offley; Morgan; Mackenzie; Smith Woodhouse; Gould Campbell, Butler; Martinez; Feuerheerd.

STYLE OR TYPE PORT OR SHERRY

Port "Type" and Sherry "Type" wines are made in this country from imported grapes. Grapes do not carry an Import Duty. The fruit is crushed here and fermented. The resultant wines give satisfaction to those whose purse-strings are foreshortened, but enjoy a glass of wine in preference to beer.

Vine Products of Kingston are the foremost producers of these wines, and they are bottled under many different labels by firms who purchase in bulk from them. "V.P." Wines are now well enough known to be called for by name, and enjoy a large and increasing sale.

Very often a clue to the producer of wine being sold under some flash label (Vino Excellencia, for instance) can be found in the words—"A genuine Vine Products Wine"— in other words it is V.P.!

Here it is only fair to clear up a slander which has prevailed too long among the ignorant and misinformed, who persist in referring to V.P. and other "Type" and "Style" wines as "Red Biddy".

Vine Products of Kingston-upon-Thames are the foremost producers of British "style" or "type" wines. They are bottled under many different labels by firms who purchase in bulk from them. V.P. and R.S.V.P. are obtainable everywhere.

V.P. wines sell cheaply, resemble their namesakes and are sound, sweet, and warming. Vine Products produce enough wine in this country to fill 25,000,000 bottles per year.

As explained on page 68 the title Port is restricted by law. Any Sherry which is not a true Spanish Sherry must bear the name of the country of origin (in letters of equal size) preceding the word Sherry—thus, ALGERIAN SHERRY. Many excellent wines masquerade under somewhat deceptive titles and this is a pity, as they cannot hope to emulate the more illustrious growths, and in many cases would be better advised to originate a generic title of their own on which to build their fame. Many first-class Empire wines might well be known by other names and thus avoid the charge of imitation.

MADEIRA, MARSALA

These are fortified wines rarely, if ever, called for in the Public-house. Madeira is from the island of that name, Marsala comes from Sicily. Both are golden in colour, rich, and sweet. They are consequently dessert wines, i.e. best consumed after a meal—not with it.

TABLE (OR DINNER) WINES

It is assumed that you will be engaged in a normal type of Public-house, where the call for Table Wines will be limited—for these wines are generally served with a meal, and at this stage you are not concerned with the catering side.

Table Wines are much weaker alcoholically than Fortified Wines, and they have no keeping qualities once they have been opened. The effect of trapping a quantity of air (air carrries many impurities) in the half-consumed bottle by corking it up again, will send the wine "off" and the deterioration soon becomes noticeable. Thus no Table Wine should ever be left "on ullage" for consumption later on. It has been done. It still is done, but a customer who is able to drink a glass of say, Claret, a week after the cork has been drawn, has no palate, probably enjoys lighting up the butt end of a cigar—and would be better advised to drink beer.

More and more Public-houses nowadays, however, do make a point of selling Table Wines "by the glass". In these cases they must enjoy a fairly rapid sale and the wine, therefore, is not left "on ullage" long enough to deteriorate.

So, if a customer walks in, and selecting a bottle of Graves Supérieure from the shelf, says he will try a glass you would be quite out of order in drawing the cork to serve him. Unless he is prepared to take and pay for the whole bottle, or half-bottle, he must be politely refused, for it may be six months before anyone else will ask for a glass of the same wine—by which time it will be turned to vinegar and represent a dead loss to the house. Even if he indicates that he will take the whole bottle it may be because he has seen the "Off-Sales" price marked up. So his drink becomes subject to "corkage".

CORKAGE

An explanation of corkage, and the reason for it, will not be out of place here, for it seems to cause raised eyebrows in many cases.

Put it simply: An apple costs you 3p in the Greengrocers. The same apple costs 10p on the bill of the Splendide Hotel. Why? Obviously because you have received some service— a plate, a fruit knife, a comfortable chair, a warm room, a waiter to serve you, kitchen staff to wash-up after you, who also had to be provided with hot water, detergent, clean overalls, cloths, etc., etc. The same thing applies to wine. Let us see what the customer expects but often resents paying for. He wants his bottle opened—he hasn't got a corkscrew! He wants his wine served in a nice wine glass—yours! He wants the glass to be polished—but he has no cloth! He wants to sit on a comfortable chair—yours! . . . at a table—yours! . . . near the fire—yours! He may spend a pleasant hour in your bar listening to the music or watching something on television. He may even ask for the loan of an evening paper . . . yours! When he has finished his wine he wants *you* to remove the empty bottle and dispose of it, wash-up his dirty glass, replace his chair and empty his ashtray! And if he accidentally breaks the glass he'll probably walk out without one word of apology.

Corkage is justified in every way—it is merely a charge for services rendered and amenities provided. So you need never feel self-conscious in making the extra charge over and above the Off-Sales price for "Corkage". The amount to be charged will depend on the rule of the house. Some establishments make a fixed amount irrespective of the off-licence price of the bottle; in others it will be a set percentage on the price of the bottle or half-bottle ordered.

TYPES OF TABLE WINE

There are so many classes of table and dinner wines that it would be only confusing to list them here. In due course you will learn as many of them as you can (wine merchants' catalogues help). The important thing, at this stage, is for you to find out all you can about the wines you *do* stock, especially which are red and which are white and, perhaps even more important, which are broadly speaking sweet and which are broadly speaking dry. If a customer asks you for a "dry white wine" and you serve him with a Sauterne, he is fully entitled to return it—and, since he has probably sipped his glass, the result is pure waste.

If a customer asks "Do you serve by the glass?" he *undoubtedly* means table or dinner wines, and if you *don't* sell these types by the glass your correct answer should be "Only Sherry, Port and Vermouth".

Table Wines are about half the strength alcoholically of Fortified Wines, and the most common types of Table Wines found in the Public-house for Off-Sales or Restaurant use are:

French White Bordeaux—(Graves, Sauternes, Barsac, etc.).

French White Burgundy—(Pouilly-Fuisse, Chablis, Hermitages, etc.).

French Red—Claret and Burgundy (Beaujolais, Beaune, St. Julien, etc.).

French Pink—Rosé (pronounced Rosay).

Italian—Chianti (Red and White, but more usually red).

(If you are uncertain whether a bottle labelled VOLNAY, for instance, is a Claret or a Burgundy,[1] the shape of the bottle gives a clue. Claret is usually bottled in upright shouldered bottles, similar to those used for Port Wine. The Burgundies are in bottles with a sloping neck, almost identical to the Champagne bottle, and have the same punt-end.)

Hock

Moselle

(Here again you may be unable to identify one from the other, but remember, Hock and Moselle wines are always in long thin narrow bottles—almost all neck and no bottle—and the Hocks are in *brown glass*—the Moselles in *green glass*.)

Chianti—White and Red (Pronounced Ki-anti). (A round bulbous bottle covered in rafia from Italy.)

The house may also carry cheaper imitations of some of the above under the label of Spanish, Portuguese, Yugoslav, etc. viz.

Spanish Sauternes, etc.

Spanish Burgundy, etc.

OPENING FORTIFIED AND TABLE WINE BOTTLES

Three other items worthy of mention crop up here concerning the opening of bottles of wine. Nearly all Fortified Wines these days have a cork with a flange at the top; which makes opening a very simple matter. *Table Wines* usually have a long cork driven right home flush with the top of the bottle, and require a *proper* corkscrew for their extraction. The thin bits of twisted wire which often masquerade as corkscrews are useless on many corks, as they do not bite into the cork properly, and, especially if it

---

[1] Claret comes from the Bordeaux area of France, Burgundy from the area of that name.

happens to be a soft one, will merely succeed in pulling out the centre—and nothing else. A *proper* corkscrew is one which has a wide cutting edge to the blade, so that it can get a real grip over a fairly wide area.

There is on the market a fine double-handled, lever action corkscrew—originally from Italy—which leaves nothing to be desired. As the screw is driven home with the aid of the "butterfly" top, the two handles rise to the horizontal position. By pressing the two handles downwards the cork is easily withdrawn from the bottle. These corkscrews will defy the hardest cork, and have the great advantage that when the cork starts to move (the critical moment of withdrawal) the hands are well away from the bottle, so the risk of injury due to the bottle breaking is negligible. They are ideal for opening Vintage or Crusted Ports when a smooth controlled action is vitally necessary.

It is not a bad idea for everyone in the Trade to secure one of these tools as his own personal property. They are well worth the money invested, and should last a lifetime.

Always drive the corkscrew well home before starting to pull. If the screw goes right through, a little chip of cork may break away and fall into the wine. If this happens do not start prodding around inside the bottle trying to get it out with a ball-point pen!—pour some of the wine into a glass and remove the chip with a spoon.

Incidentally, here you have one of the reasons for a sample of wine being handed to the host before guests are served at table. This is not just a piece of flamboyant showmanship (ten to one the host could not tell a Liebfraumilch Blue Nun 1955 from a Rudesheimer Deinhard 1959 unless he saw the label). It is done merely to ensure that nothing like our little chip of cork will be served to any of his guests, and that the wine itself is not corky, sour, or otherwise undrinkable.

The next point concerns the very soft and sodden cork which cannot take the bite of an ordinary corkscrew, or perhaps disappears straight inside the bottle. Take a piece of fine twine (not rope), make a loop in it and push the loop into the bottle. By turning the bottle on its side endeavour to get the loop round the bottom of the cork after manoeuvring it as near the neck as possible. It may take a minute or two of juggling, but eventually you will have the twine round the base of the cork, and by pulling the two ends, be able to withdraw it.

The third point in opening a bottle of wine concerns the metal capsule. Any metal is bad for wine, and should not come in contact with it—even metal wine measures are not good—but are frequently used. Therefore, when opening a new bottle of wine remove *ALL* the metal capsule from both the cork *and* the bottle. With a clean cloth wipe the dust from the lip of the bottle—every time.

SPARKLING WINES

Sparkling "Made" wines include Champagne, and it may fall on you to serve this in your bar on rare occasions.

"Made" wines are those which have been treated to special processes additional to the normal fermentation.

There are Vintage years with Champagne, and this appears on the label. This means that the wine was reputedly the product of that particular year. Non-vintage Champagne, which incidentally may sometimes be particularly good, is a blend of various wines and the wines of various years. The name Champagne is jealously guarded. The district in which it is grown is very small and only wine produced within the exact limits of this area is entitled to be called Champagne— by French law. The word Champagne must appear on the cork and on the label. It is shipped under the name of the grower thus:

"Bollinger 1955" or

"Pol Roger—Non Vintage" (Sometimes just N.V.)

Any of the following terms may also appear on the label:

Brut Nature Extra Dry Extra Sec Très Sec	all meaning Dry
Dry Sec	meaning Medium Dry
Demi-sec	meaning Medium Sweet
Demi-doux Doux Rich	all meaning Sweet

but, one other name will appear on the label—Rheims or

Epernay. These are two centres of the Champagne area from which the wine is shipped.

Well-known shippers include the following names: Ayala; Bollinger; Clicquot-Ponsardin; Goulet; Chas. Heidsieck; Heidsieck et Cie; Irroy; Krug; Pol Roger; G. H. Mumm; Lanson; Moet et Chandon; Perrier-Jouet; Delbeck; Piper-Heidsieck; Pommery et Greno; Lemoine; Deutz et Gelderman; Binet; Louis Roederer; Ruinart; Duminy.

It is very unusual for Champagne not to be served iced—about as unusual as trying to clean brown shoes with black polish! In view of this, any order for Champagne must be taken in plenty of time to allow for the chilling. Ice may be packed round the bottle (or bottles) in an ice bucket, if one is available, otherwise in an ordinary bucket—kept well out of sight of the customer. However, iced water is to be preferred as it does not make the wine "patchy" and gives it a more even chill. Twenty minutes will be found long enough for this procedure.

For Public-house purposes Champagne sizes are bottles, half-bottles and sometimes quarter-bottles. It is often known as "Bubbly".

Black Velvet

Whilst on the subject of Champagne, it is just possible that some time or other you may be asked to serve "Black Velvet"—even if it never comes your way, as a qualified bartender you should know what it is and how to concoct it. Although this may mean one of several things, the *real* "Black Velvet" is Guinness and Champagne.

Curiously enough, if called for at all, it is usually in the morning, by wealthy car dealers, bookmakers, or publicans out on a spree, all smoking frayed cigars and full of their own importance. There is, of course, no accounting for tastes, but this drink may be safely labelled as an example of pure exhibitionism. Guinness is a great drink. Champagne is a vinous jewel. Taken as they were intended, they are the very best of their respective classes, and it always seems questionable which of the two is suffering the adulteration in "Black Velvet".

However, it is important to make some enquiries right at the outset, the first of which will be whether the customer requires Guinness and Champagne (he may want Mackeson and Moussec—or something else!). Secondly—how many persons require it? Does he want a bottle of Champagne—

or a half-bottle? Any particular Champagne—vintage or non-vintage (providing you keep them). Does he require the Champagne iced? Will he wait while it *is* iced? (The class of person who drinks this may not be averse from having a lump of ice in the glass!)

NOTE: Should it happen that the call is a frequent occurrence, it is quite likely that Champagne is on ice for this purpose. Enquire.

Having secured all the desired information, and remembering that this drink is usually called for to make an impression on the assembled company, the bartender should prepare the concoction with a certain amount of pomp and circumstance. He, or she, should polish the glasses vigorously with a clean cloth—even though they have just been done! Then place them on the counter in front of the customer (his back will be turned, anyway!). The glasses should be twelve-ounce ones—the best stem type available. With regard to the actual pouring—*the Champagne goes in last*. The usual proportion is half and half—but the customer sometimes prefers to pour in the Champagne himself—that is his privilege.

Opening Champagne (or any Sparkling Wine)

A cloth is always wrapped round the bottle for three valid reasons—and hiding the label is *not* one of them!

Firstly, the bottle is generally dripping wet from the ice tub.

Secondly, through being wet it can easily slip through the hands, leaving the bottle on the floor, and you holding the label.

Thirdly, when the cork is drawn the wine often bubbles over—*should* bubble over, in fact!

Due to the pressure exerted by the secondary fermentation inside the bottle (which makes the bubbles) Champagne is always bottled in extra heavy bottles, and the corks have to be wired on to prevent them shooting out. So having wrapped the bottle in a clean napkin, leaving the neck free, remove *all* the metal foil from the cork. Then pointing the bottle upwards and not towards anyone, carefully unwind the wire. Sometimes the cork will shoot out as soon as the wire is loosened, but often it will hold until it is nearly drawn. There is always a loud "pop" which turns the heads of everyone within earshot, bringing forth cheerful chatter and admiring glances, much appreciated by the man in the money. Do not worry too much if the wine spurts up in the air like a fountain. Do not panic! Do not put your finger over the neck—that will only make it worse! Do not point it at anyone and give them a showerbath! Take a corner of the napkin and hold it over the opening for a second or two. This will allow the excess gas to escape. You will not have lost much of the actual wine in any case, for the fountain is all gas and froth, even though it may have made quite a mess on the floor. But *always* be ready for this outburst with your napkin.

Just before you actually open the Champagne, you will have half-filled each glass with Guinness, very steadily poured, so as not to have shown too much "head"—the less "head" the better, in fact. Then enquire of the customer if he approves of the quantity of stout in each glass, and perform all this with the Champagne—iced or otherwise—by your side ready to open. Top up the glasses with the "Bubbly".

To make quite certain of the procedure, it is given again in proper sequence:

1. Enquire what sort of Black Velvet is required.
2. Enquire what kind of Champagne—N.V. or Vintage.
3. Enquire if bottle or half-bottle is required.
4. Enquire if "Bubbly" is required iced.
5. Enquire how many people are to drink it.
6. Secure best stem glasses.
7. Half fill glasses with stout.
8. Open the Champagne and top up the glasses.
9. Do not panic—whatever happens!

BRITISH SPARKLING WINE
Moussec (meaning—"Sparkling Dry")

This dainty and pleasant little drink was the first to be bottled in a "baby" size. It is the original baby wine. It must in no way be confused with "Babycham" which is not a wine but a Perry.

Moussec is made from French grapes, the juice from which is processed in France and specially imported for the purpose. The juice is fermented with a special yeast, and the wine is bottled in Hertfordshire. The sparkling bubbles in the wine are entirely natural and are produced through what is known as double fermentation as is the case with Champagne.

Moussec is not a Champagne, does not set out to be Champagne, nor is it a sham Champagne! It is a British Sparkling Wine. The fact that it might be mistaken for champagne is a tribute to its quality. It offers very good value for money in view of the trouble and care taken to produce it, and for a lady unable to make up her mind (as you will find is often the case) it may safely be recommended. There are two kinds—sweet and dry. The dry variety has a medallion in red on the neck-band and the sweet has the same in blue.

No decoration is requested by way of a cherry (unless, of course, it is specifically requested, and then you have the measure of the customer's ignorance and bad taste!). It is, however, vastly improved by being served in a footed Champagne goblet and this also has the additional advantage of giving the lady a bit of "one-up". Some customers like it with brandy—an admixture of advantage to neither product but that will be none of your business. The term "Cuvée Nature" on the neck label means that the wine is blended to give a dry finish.

It is very suitable for making "Black Velvet" in place of Champagne. It is much cheaper, and it has the additional advantage of being bottled in very handy sizes for this purpose:

> Baby size—with crown cork
> Bon Santé (¼ bottle) with crown cork
> Half-bottle
> Bottle

NOTE: Bon Santé is pronounced Bon-Son-tay.

Merrydown Apple Vintage Wine

This has been dealt with under the heading of Cider and Perry in Part I (Merrydown) for although it is now known as Apple Vintage Wine, it is in fact actually a high strength Vintage Cider.

BRITISH STILL WINES, GINGER WINE, etc.

Ginger Wine, much used in "Whisky Mac" (the making up of which see under Whisky, page 49), is also a sweet and warming drink by itself. Stone's, much advertised, is a favourite and no complaints will ever be received when this is served. It has one great virtue—it is consistent and no wonder for Stone's have been marketing British Wines for over 200 years! Some Ginger Wines are very "gingery" and hot, others do not seem warming enough, being more syrupy. This, of course, is for your information—you will have to serve what is provided for you, but bear it in mind for the future. Other British Wines which may, or may not, be found in Off-Licences are: Orange, Raisin, Peach, Damson, Elderberry, Cowslip, Apricot, Cherry and Blackcurrant etc., but they are not usual Public-house lines for consumption "ON" the premises.

VERMOUTH

There are many brands of Vermouth, both from Italy and France—the French varieties being lighter and drier than the Italian, which are generally sweet.

A white wine is used to produce Vermouth to which various herbs, extracts and aromatic ingredients are added. Although they are sometimes consumed "neat" (exactly as poured from the bottle) or as a long drink mixed with soda and ice, Vermouths are usually served mixed with Gin (or nowadays, Vodka).

Martini Sweet is the most popular of the Italian Vermouths in most Public-houses, and will always be accepted when serving gin and 'IT'—for this is what the "IT" stands for —ITalian Vermouth!

About the same quantity of "It" as Gin is the usual measure—*the gin in the glass last*—and serve it with a cherry on a cocktail stick. *Ice first,* if called for.

Noilly (pronounced NOI-EY) is the most popular of the French Vermouths, and should always be used when serving Gin and "French"—and this is what "French" means— French Vermouth.

About the same quantity of "French" as Gin is usually sufficient—*the gin in the glass last*—and serve it with a piece of lemon on a cocktail stick.

Cinzano (pronounced CHIN-ZAHNO) may be obtained as French, Italian or Bianco and may be preferred by some to the Martini or Noilly (if it is stocked).

(Bianco is a light sweet Vermouth).

Two other Vermouths may sometimes be found on sale in pubs—Gancia and Brega Rossi.

The Licensed Trade suffers greatly from imitation by firms within it. Because Martini Sweet has a colourful red label, almost every Sweet Italian Vermouth has a similar sort of label. Because Noilly has a green label, then almost every other *dry* Vermouth has a green label. But this has one advantage—it is not so easy for you to mistake one for the other!

Martini and Cinzano both make a dry French.

A digression may be made here on the subject of cocktails and mixed drinks. Cocktails are not given in this manual—they are usually merely a catalogue of recipes and can be obtained in booklet form from many firms in the Trade free of charge. However, a word of warning! Do not be led too freely by these little booklets, or by directions given on bottle labels, for the publishers always have an axe to grind. Taking Dry Martini as an example, we sometimes find a house selling gin recommending it to be made with *two-thirds* of their gin, *one-third* of French Vermouth. A house selling Vermouth recommends *one-third* of Gin and *two-thirds* of their French Vermouth, whilst a cocktail recipe book may well state *one-half* of each.

One recipe is given here, often requested by American visitors, and usually refused because nobody in the bar knows how to make it—the long drink called Tom Collins.

> Put 3 or 4 cubes of ice in a stem glass. Add the juice of one lemon, two teaspoons of sugar. Add a measure of (or a double of) Gin, and fill up with Soda Water. Stir it quickly.

Too easy!

(A *John* Collins is the same recipe using Hollands Gin).

SERVING WINE

No wine should ever be served in a glass which does not accommodate it comfortably—with room to spare. In many houses wine is served in little glasses brimful to the top. This makes their removal from the counter a hazard to clothing, and does not allow for a proper appreciation of the drink itself.

The customers themselves are often to blame for this, because seeing a glass only half full they think they have been "done" and want it "topped up" for the same money. Although they might actually be getting less wine, a little glass full up seems like full measure. Funny people, the public—and where drink is concerned funnier still, you will learn.

It goes without saying that all wine glasses must be

absolutely clean and POLISHED. In any case always give them a quick shine with a clean cloth before you serve, especially if your wine trade is not very brisk, because they may have become dusty on your shelves.

THE TONIC PROPERTIES OF WINE

The public is generally not aware of the tonic properties of wine. Beguiled by advertisements on hoardings and T.V., they willingly pay a price for a Tonic Wine which promises so much. The following is a guide which may be used to advise customers who require a wine to perform a specific service, and to act more as a medicinal tonic than as an embellishment to sociability.

Loss of appetite: A *dry* Sherry will often give an edge to a jaded appetite if consumed shortly before a meal, especially after illness.

Anaemia: A large glass of Burgundy twice a day at, say, 11 a.m. and 9 p.m. perhaps with a biscuit is a wonderful tonic for those whose blood is lacking.

After the 'flu: Ruby Port of good quality is as good as anything to bring back rosy cheeks and life to those who feel the need for "something" after this, or any other debilitating illness.

Rheumatism, the scourge of so many folks not always elderly but old before their time by reason of the pain which offers such excruciating agony.

Long-term preventive measures are not so popular with those who are suffering right now, but it should be impressed upon them that *fermented* liquors are bad for the complaint. Fermented liquors are, of course, beers, ciders, and all wines. Alcohol itself, it has been found, has no detrimental effect, but the manner in which it is produced certainly does.

Thus the only drinks available to the rheumaticky are those which have been distilled—the spirits—Gin, Whisky, Rum, Brandy and Vodka.

Livener: Champagne—not necessarily an expensive vintage growth—will work wonders in reviving the flagging spirits, restoring energy and fighting the "blues". Unfortunately it is expensive, and because of this is reserved mainly for celebrations, but if you have a customer who badly needs pulling out of a fit of depression (and can afford the money!) even a quarter bottle of Champagne could do the trick. As an alternative let him try Moussec (Bon Santé) for a similar effect.

Digestion: The alcohol in wine aids the digestion. A glass of wine with, or after meals assists the digestive juices to do their work to better advantage—especially in the elderly.

Taken all in all wine, the unadulterated and naturally fermented juice of the grape, is a great giver of health, and most people would benefit by drinking more of it.

8
ALCOHOLIC STRENGTHS OF
WINES AND SPIRITS

A gentleman named Sykes, who was an officer in the Customs and Excise Department over 150 years ago, perfected an instrument called a Hydrometer for the purpose of measuring the strength of alcoholic spirits. It is still used today and you may see one operated in your bar. Although this may be a sign that there is some doubt about the spirits on sale, it is not always so, as some companies have a periodic purely routine check made of the spirits exposed for sale in their houses by outside inspectors.

If you look around your shelves and examine your Whisky and Gin bottles, you will see one or the other of these strengths printed on the label:

70° Proof Spirit

30° Under Proof

They mean the same thing. The strength of spirits given as either of these (70° Proof or 30° Under Proof) is arrived at as follows:

PROOF Spirit equals 100°. It is made up of approximately 57 per cent pure alcohol and 43 per cent *liquor* (by volume). (The *Liquor* contains the by-products of the item from which the spirit is distilled—in the case of Brandy it is wine, in the case of Whisky it is a mash of barley grains, etc., in the case of Rum it is molasses.)

So it should be clear that 100° Proof Spirit is *not* 100 per cent pure alcohol—it contains only 57 per cent alcohol.

Having therefore arrived at the alcoholic strength of *Proof Spirit* as 100 degrees, we now require it to be 30° *under* Proof, so thirty parts of the proof spirit are removed, and thirty parts of distilled water are added. The bottle now contains 40 per cent alcohol and 60 per cent liquor, and is said to be 70° Proof, or 30° under Proof.

Now it is possible to distil and re-distil until every particle of the original product has been distilled away, then all that is left is absolute alcohol, flavourless, colourless and mighty potent—but no good for drinking purposes.

As Proof Spirit contains only 57 per cent alcohol it is possible to have spirit *over* Proof, i.e. 110° Proof or 10° *over* Proof.

This is something which the general public often find difficult to understand, because they *will* confuse 100° Proof Spirit with 100 per cent absolute alcohol. Sometimes it is very difficult to convince them that it *is* possible to obtain spirits at 110° Proof. Rum, for instance, arrives in this country at 35° *over* Proof or 135° Proof, and there is one drink (if that is what you dare to call it!) known as POLISH PURE SPIRIT which is on sale here at 140° *Proof*. To size up:

One Hundred Degrees Proof Spirit (100°) contains 57 per cent Alcohol and 43 per cent by-products of the item being distilled.

Seventy Degrees Proof Spirit (70° proof) or ⎱ the same
Thirty Degrees *Under* Proof (30° U.P.) ⎰ thing
is obtained by removing 30 parts of *Proof Spirit* and adding 30 parts of distilled water. It now contains 40 per cent Alcohol and 60 per cent by-products and water.

As the Duty on spirits is based on the alcoholic strength, the standard of 70° Proof (30° U.P.) is generally accepted as a compromise between a reasonable strength for public consumption and paying the full Duty, as it therefore only attracts 70 per cent of the full rate, which means you can buy it more cheaply, but you can't get "stewed" so quickly!

The belated panic to push our way into the European Common Market might have had some peculiar side-effects on the Licensed Trade, although at the time of writing the results have not yet brought major changes.

The incongruity of a navvy asking for a litre of Bitter and deux grams of shag until he is converted to smoking Dutch cigars and shipping Pernod will be worth staying to watch the second time round.

One effect, well overdue, will be the adoption of the Gay Lussac measurement of alcoholic strengths instead of our present complicated and antiquated system based on Sykes.

In the Gay Lussac system water is zero and absolute alcohol is 100. The strength of spirits is denoted by the amount of ALCOHOL it contains—not some admixture of water and alcohol little understood by the layman.

Thus 50 per cent alcohol and 50 per cent water is marked as 50° Gay Lussac. It is *that* simple. So our Scotch whisky at

70° proof which, as explained before, actually contains 60 per cent water and 40 per cent alcohol will be known as 40° Gay Lussac.

Very shortly you will begin to see spirits marked in both methods on your shelves and you will be able to explain to your customers that they are one and the same as regards the alcoholic content—even though they may half suspect that they are being swindled by what *looks like* the weaker 40° Gay Lussac.

One point should be made clear here. You may hear someone talking about the strength of American spirits. For your information they are *not* the same as "Sykes strengths" —they are, in fact, considerably weaker.

70° Proof here equals 80° Proof in U.S.A.

100° (Proof Spirit) here is reckoned to be 114° in the U.S.A.

American whisky marked 86.8° Proof would be equivalent to 76° Proof here, and their 100° (Proof Spirit) is only our 87.6° Proof, by our standard nearly 12½° weaker.

In the case of Wine, strengths follow the same standard except that instead of water being added to break the strength down alcohol is added to bring the strength up, as was explained under the heading of Fortified Wines.

On some bottles of wine will be seen: N.E. 25° Proof Spirit.

This means *not exceeding* 25° Proof Spirit—i.e. it is 75° *under proof,* and is so stated on the label for Customs and Excise Duty.

Port and Sherry come in at about 35° Proof Spirit (N.E. 35°). It is therefore half the strength of Whisky at 70° Proof Spirit.

9
SPIRITS

WHISKY

Scotch Whisky

There are two kinds of Scotch Whisky—Malt and Grain. There are two methods of distillation—the Pot Still and the Coffey Patent Still.

Scotland produces the finest barleys in the world and this combined with the soft and gentle waters from the moors and Highland streams is the reason for Scotch Whisky being acclaimed the very finest. Every whisky made outside Scotland tries to emulate the home product. The combination of soil and climate is not transferable—the forces of nature do not react willingly to interference.

Thus the delicate aroma of Havana leaf produces the finest cigars and transplanting the tobacco weed anywhere else in the world, under identical conditions, still fails to produce anything comparable to the Cuban growth. Cognac Brandy is the finest in the world, because of the soil and climate producing the particular wine from which it is distilled. It cannot be improved upon, nor can anything approaching it be made anywhere else.

So with Scotch Whisky. Whisky is made in many parts of the world—including Japan!—but Scotch Whisky stands alone, placed on a pinnacle by Nature—not by design. Not that man by his care and patience is to be underrated, for given the finest materials to work upon, it is only by his experience and consummate skill that the peak of perfection is attained.

Malt Whisky is made from malted barley only, grown in selected districts of Scotland, both in the Highlands and the Lowlands, and it is made in the old-fashioned pot still. Malt Whisky is full flavoured and full bodied, and expensive, but is not acceptable as a drink on its own to everybody.

Grain Whisky is a much more commercial spirit produced

90

by the patent method, and is made from maize, oats, rye and a proportion of malted barley. This process is much more rapid and a more neutral spirit is obtainable with little flavour.

Malt Whisky takes much longer to mature than Grain Whisky, but it is not found palatable by the public. It is therefore blended with Grain spirit to a standard consistent with consumer demand. It is the Malts which give Whisky its delightful flavour.

When distilled, whisky is the same colour as gin, but it is coloured artificially to suit the individual brand. (A colourless whisky is on the market for use by diabetics. It is sugar-free.)

NOTE: The Distillers Company, known in the trade as DCL, is an amalgamation of distillers founded nearly 90 years ago, and controls a vast number of brands of Scotch Whisky.

The largest selling Scotch Whisky in the world is Johnnie Walker (DCL). The largest in this country is John Haig (DCL).

It is just one of those things, but you will probably notice that Haig is tops in the West-End and better class bars, whilst Johnnie Walker sells better in the East-End, public bars and more earthy pubs.

Whisky Mac

This is Whisky (Scotch) and Ginger *Wine* mixed. About the same quantity of each is used. No importance is attached to which goes in the glass first, as they blend together very well. For a cheaper "Whisky Mac" Ginger *Cordial* is used, but this has little to recommend it—apart from a copper or two on the price—as being non-alcoholic it has the effect of diluting the Whisky. *When serving Whisky Mac with Ginger* Cordial *the whisky goes in last.*

IRISH WHISKEY

This is by common usage spelt with an "E"—Whiskey. It is manufactured from barley and other cereals in the old-fashioned "Pot Still". It enjoys a flavour quite distinct from Scotch Whisky, due in part to the peat used in the production which gives it a decidedly "smoky" taste. Curiously it is not so popular as Scotch—even with Irish customers—not in this country at any rate.

RYE WHISKY

As the name implies, Rye Whisky is produced from Rye in America and Canada, and is distinct from Scotch in flavour and finish—being harsher.

Bourbon Whisky, widely advertised in American magazines, originates from Kentucky.

GIN

London Gin

Gin is the purest of all spirits and is ready for consumption immediately it is distilled (with one very notable exception—see under "Booth's Gin" below).

It is not left to mature in casks as is the case with brandy, rum or whisky.

Grain spirit, from which most gin is made, is distilled principally from maize, which is then re-distilled (a process known as rectifying). It is then flavoured by an essence made from juniper berries. Other items used by various "rectifiers", i.e. corriander seeds, cinnamon, Angelica, carraway and other aromatic roots, are the secret of each.

Other gins are derived from molasses, but grain spirit is far superior. Some gins fell into a decline during World War II when the use of grain was prohibited for practically everything—even chickens! The Government issued the spirit to the rectifiers—and, like many amateurs, didn't know bad from good—with the result that one or two unfortunate firms secured a bad name with the public, which they found difficult to shake off afterwards. However, there is no bad gin now, although there is considerable variation in flavour and palate between them—some being considerably less pungent than others—although of the same alcoholic strength.

The word "Gin" is a corruption of the word "Geneva" —not the Swiss city of that name but from the old French "genevre" meaning "juniper". The word Geneva appears on the labels of Hollands Gin bottles.

Booth's Gin

Gins which are golden in colour are said to be "straw-tinted". Booth's was the first gin on the market to be sold coloured—and this happened by accident!

Many years ago there was over-production at the distillery

and a storage crisis developed. All that could be found as containers for the excess was a batch of casks in which Sherry had been shipped to London. The gin was poured into them and they were set aside.

Eventually, when broached the gin was found to be most attractively straw-coloured. Tasting showed that, far from having any serious defect, the gin was considered to be mellow, and when sold it immediately found favour with gin drinkers. What had been forced upon the firm by circumstances then became standard practice and all Booth's Finest London Dry Gin was subsequently rested in wood.

Plymouth Gin

This gin has a flavour quite distinct from that of the London Gins. The water used in making Plymouth is remarkably soft and runs through and from the moors of Devon. It is this which helps to make it different. It has always been a great favourite with officers of the Royal Navy and is most suitable for making "Pink" Gin (see page 123).

NOTE: The sale of Plymouth Gin may be increased considerably if it is given prominence on the display cabinet. Many people who would be pleased to have it show a reluctance to ask for items which are not shown in case they are not stocked.

Squires London Dry Gin; Cornhill London Dry Gin

These are Gins sponsored and marketed by combined Brewery companies. In other words they are the Brewers' own brands of Gin. For this reason they are expected to be given prominence on display shelves, cabinets and dispensers. Inspectors and other gentlemen from the Brewery will want to see these two lines given the premier service position in your bar and not only that!—they will expect you to SELL them, too!

A house under the control of the sponsoring Breweries which shows a good trade in these Gins earns for itself a commendation.

Squires is a white Gin. Cornhill is straw-tinted.

Hollands Gin

Produced in Holland, this is a Gin with well recommended medicinal properties, especially in disorders of a rheumatic or gouty nature and also for kidney troubles.

It is very much more pungent than other Gins and is

usually taken as a medicine. It should be kept on ice and served cold.

The name "Geneva" appears on the label of most brands of Hollands Gin, and as explained before means "juniper".

Pink Gin

This is Gin to which has been added a very small quantity of Angostura Bitters. It is made up behind the bar as required in the following way:

Shake from six to ten drops of Angostura Bitters into the glass (these bitters have a shaker top supplied with each size bottle). Roll the glass in both hands to ensure the bitters cover as large an area of the inside surface as possible, then shake what is left in the well to dispose of the surplus. Dispense the measure of gin called for into the glass. The amount of bitters still adhering to the glass is sufficient to tint the gin a delicate rose pink.

Any Gin will make "Pink Gin" but there is, and always has been, a bias in favour of using Plymouth Gin for this purpose—and this is often called a "Pink Plymouth". You would never be wrong to serve Plymouth Gin when a Pink Gin is called for.

NOTE: Angostura Bitters because of its use in making up Pink Gin is itself known as "Pink"—so if anyone asks for the "Pink" you will know they are referring to Augostura Bitters.

GIN CORDIALS

Gin, being particularly pure due to the rectification of the spirit, combines most favourably with many other flavourings or distillates, chief of which is the sloe (the fruit of the blackthorn), at one time quite a popular drink. Both Damson Gin and Sloe Gin are hardly every heard of nowadays.

GIN SLING (PIMM'S NO. 1 CUP)

The only Gin Sling likely to concern you comes in a bottle and is known as Pimm's No. 1 Cup. It has gin as a base, and contains various other secret ingredients. It is a very pleasant and popular drink—especially with the ladies. It is often, but improperly, served looking like Covent Garden market —the glass being stuffed with apple slices, orange portions, lemon, cucumber, cherries, sprigs of mint—and anything else handy. This may delight certain young ladies, but as a drink and not as a dinner it is overdoing the "fruit and veg" a bit.

The Pimms bottle is graduated into sevenths—each one estimated to be sufficient to make a pint when filled up with lemonade, but you will be guided by the rule of the house, both as to the method of making up and the measure to be used.

Pimms suggest a piece of ice, a piece of lemon and a piece of cucumber rind or a sprig of borage (pronounced burridge) as the only garnish the drink requires, but it is generally agreed that the fruit decoration, if not overdone, helps to sell this drink. (Borage is not always so easy to obtain so the proprietors are prepared to supply packets of borage seeds to anyone interested enough to grow this herb themselves.)

Mugs (with handles) are **always** used when making up the larger size of Pimms. However, there is a growing tendency to serve a small Pimms, known as a "Pimlet"—using a 6-out measure of Pimms instead of the 3-out (double)—and in this case a large spirit glass is used. It sells for half the price of the half-pint and is quite a popular size.

Gin and Lime (Lime Juice Cordial), **Gin and Orange** (Orange Squash), **Gin and Lemon** (Lemon Squash), **Gin and Pep** (Peppermint Cordial).

Owing to the fact that the gin in all these drinks is a constant factor they have been dealt with under the heading of *Soft Drinks*—as it is the Squashes and Cordials that are variable (see pages 42-8, Part I).

RUM

Rum is made from molasses which comes from the sugar cane. The sugar cane grows profusely all over the West Indies, particularly in Jamaica, Barbados, Trindad, and also in Demerara in British Guiana.

Molasses is the residue left after the crushing of the sugar cane. It is fermented and then distilled to a very high alcoholic strength. It is at the raw sugar production points that Rum is produced from the molasses which are a by-product of raw sugar-making and Rum can only be produced at this source. Rum has no connection at all with the re-fining of sugar.

Rum from each area has a very distinctive flavour which makes it impossible to substitute one for the other—so don't try it! Jamaica Rum is usually a golden brown colour, soft and delicate, but colouring may be effected by the addition of a caramel substance made from burnt sugar, as is the case with some of the very dark rums.

Rum improves greatly with age where the "fire" is absorbed into the wood of the casks. It does not age in the bottle. It is said that in London's Millwall Docks are casks of Rum laid down in the time of Nelson, which should by now be equal to a fine liqueur—or else be empty—if we know the docks!

Rum has close connections with the sea and nearly every label indicates this. Formerly it was a daily issue in the Royal Navy, but now sailors have a cash allowance in lieu.

Caroni Rum, a Trinidad Rum produced by Tate and Lyle, is on the market here at two strengths, 75° and 90° Proof, and is quite different from the usual rum in that it is not so pungent in flavour and, in fact, to many people does not savour of rum at all. When you find a bottle on ullage do not be surprised if it does not smell like rum, or at least not as you would expect rum to be for it is far more delicate in bouquet and more gentle on the palate. Caroni Rum has for many years been sold to the Admiralty who are in fact its biggest single customer, but other rums are also purchased to make up the Royal Navy blend. Caroni is called Navy Rum—a title which is not restricted.

The subject must not be left without mention of the products of United Rum Merchants which you will almost certainly have exposed for sale in your bar—probably on the "optic" as they are all popular and fast-selling Rums:

Lemon Hart Rum—Jamaica
Lamb's Navy Rum—Demerara
Santigo—Light Demerara

Of these the famous LEMON HART is a great favourite amongst the Jamaican Rums and it is a splendid base when mixed with "Black" (Blackcurrant) or "Pep" (Peppermint), but its enchantment is perhaps best appreciated when the delicate flavour is not dealt with quite so harshly but is mixed with something softer—"Orange" (Orange Squash) for instance, for then it can scarcely be bettered.

Lamb's Navy Rum is a traditional dark Demerara Rum and is particularly popular round the coasts and in sea-port towns. It is a good mixer but is frequently drunk "neat". The brand enjoys an enormous export sale and whilst not as old as LEMON HART RUM—which first found its way on to the market in 1804—it is probably the biggest selling brand of Demerara Rum on the market today.

Daiquiri cocktails need white rum; these are light in colour, light in palate, light on the nose. A perfect mixer for long summer drinks or as a main constituent of Rum Punch in the winter. This is quite a new departure in rum marketing following a trend towards a lighter rum which is so marked in America and Canada and, of course, also in the West Indies.

Among the rums you will doubtless come across one called *Bacardi*. This is the best-known light rum and is extensively called for as a base for cocktails when the strong flavour of the ordinary rums would not be acceptable.

"Neaters": A name by which Navy "issue" rum was some-known, also "P.O.'s Neaters".

"Nelson's Blood": Another slang term for rum.

Tom Thumb: Cockney rhyming slang for rum.

Grog: Sometimes used to describe rum. Again the name has maritime connections as Navy "issue" Rum (diluted) was introduced by Admiral Vernon in 1745. The Admiral used to wear a coat made of a coarse rough material known in those days as grogram and thus attracted to himself the name of "Old Grog". Later the name devolved upon the drink itself and then to any kind of spirit.

BRANDY

If brandy is the Prince of Spirits, then Cognac brandy is the Queen—majestic, imperial and seated on a throne way above the heads of all others, a throne which will never fall.

Unlike other spirits brandy is made from Wine—distilled and matured to produce a drink of great charm and medicinal merit. Any country producing wine is able to produce a brandy of sorts, and many of them, notably Spain, Portugal, Italy and Southern France, distil great quantities.

But it is from the small and strictly delineated area on the River Charente in western France that Cognac draws both its name and its fame. No other Brandy may bear the name Cognac unless it originates from wine grown here. (Cognac is pronounced CO-NIACK.)

Cognac itself is a very small part of the brandy-producing area, but within this district are distilled the famous Grande Champagne, Petite Champagne and Borderies—all brandies of outstanding merit and nothing whatever to do with the sparkling white wine of Rheims and Epernay—known as Champagne.

Next in quality to Cognac comes another fine Brandy known as Armagnac. (Pronounced *AR*-MAN-NIACK). This is produced in a very small quantity in the Department of Gers in S.W. France. It is highly esteemed and may be found in the better-class public-house.

The wine from which Cognac Brandy is made is thin, acid and quite undrinkable, but after distillation (in the old-fashioned pot still) it has absolutely no equal. Brandy improves vastly in cask, leaving its fire in the wooden staves and taking on a mellow maturity during the years. Vast quantities are stored in Cognac and the surrounding districts patiently awaiting the soothing touch of time. The House of Martell alone place the conservative estimate of ninety thousand hogsheads in stocks held in their warehouses.

The very finest Grande Champagne and Petite Champagne may be kept upwards of fifty years in cask, after which they are bottled or removed to glass containers to preserve their strength and save what is left of them, as the evaporation is considerable whilst in the wood.

It should be made clear that brandy does *not* age in bottle. For instance, brandy of 1878 bottled in 1938 is sixty years old. In 1958 it was still 60 years old, and in 1978 it will still be 60 years old.

A customer may mention that a bottle of brandy he has had in a cupboard under the stairs for thirty years must be good! Doubtless it is worse than when it was put there, owing to deterioration of the cork—and it may not even have been very good brandy in the first place.

Old Cognac Brandy is considered the finest liqueur in the world, and there is just a possibility that you may have a bottle standing on your shelves somewhere.

Amongst the best known names in England are: Martell; Hennessy; Otard; Courvoisier; Bisquit Dubouche; Croizet; Sauvion; Hine; Remy Martin; Denis Mounie; Rouyer; Salignac; Castillon.

With regard to the stars and other marks on brandy labels

these are no indication as to the maturity of the contents and consequently it is better to trust the veracity of the shipper in his own description. In the case of Martell the familiar blue label should be well in evidence in your bar with Hennessy's Bras Arme accompanying it. With these there will probably be Martell's Cordon Bleu, an old Cognac for the appreciative drinker, and Hennessy X.O. for other connoisseurs.

Good brandy should always be served in a large glass— a proper "brandy balloon", if any are available, for this drink is greatly improved by being "hand-warm". You will see your brandy expert cupping his hands around his glass, for the warmth thus generated releases the "ethers" in the spirit adding to the bouquet and to the pleasure.

Now, because of its stimulating and medicinal properties, brandy is often prescribed by doctors for elderly folk with a condition of the heart and other disorders. Here, however, is a legal point. Nothing, repeat *nothing*, in law, may be supplied out of permitted hours, even on a doctor's certificate.

It is always somewhat doubtful whether brandy, asked for at the side door "Out of hours" for someone who is having a heart attack, will do the patient any good. If a doctor prescribes brandy it is not usually in an emergency (which is something he can deal with in other ways) but as a stimulant for general purposes—and therefore the matter can wait until "Opening time". The anxiety shown by some people when seeking "emergency" brandy is, more often than not, a reflection on their own improvidence in not keeping a supply in the house.

In any case the Law makes no provision for "emergencies" in the case of Licensed premises, doctors or no doctors, so people requiring alcoholic stimulants must be careful not to be taken bad "out of hours".

VODKA

This colourless, flavourless spirit has become more popular recently due, in part, to a persistent sales push, advertising, and to a desire for certain classes to be "with it".

Vodka has been made in Russia for over 700 years, but it probably originated in Poland. The Russian and Polish products are certainly not tasteless, but have various herbal extracts and spices infused into them. Only the British product is "neutral".

British Vodka with attractive bottles, heraldic labels embellished with weird names, give this spirit an eye-catching appeal and it is now being used freely for mixing "Bloody Mary".

It sells for round about the same price as gin (Off Sales) but is often dearer in the pub—by the "nip". You will certainly be asked for it—more and more in the future if the present trend continues—mixed with Orange (a favourite) or Lime, any of the "Mixers", Vermouths or Aperitifs—or, as drunk in the Kremlin, neat.

10
BOTTLES AND MEASURES FOR WINES AND SPIRITS

MEASURES, SIZES AND QUANTITIES

The usual bottle of Whisky, Gin, Rum and Wine contains one-sixth of a gallon. It is called a *Reputed Quart*.
(These are not to be confused with Standard Imperial Measure where the quart goes *four* to the gallon.)

One Imperial gallon = 160 fluid ounces.
One Imperial quart = 40 fluid ounces.
One Imperial pint = 20 fluid ounces.
One "Reputed Quart" ($\frac{1}{6}$ of 160 ounces) = $26\frac{2}{3}$ fl. ozs.
One "Reputed Pint" ($\frac{1}{12}$ of 160 ounces) = $13\frac{1}{3}$ fl. ozs.
One gill or quartern = $\frac{1}{4}$ pint or 5 fluid ounces.
There are $5\frac{1}{3}$ "gills" or "quarterns" to the "Reputed Quart".

Spirit measures in the public-house are usually the "6-out"—(called a "single", a "nip", or a "tot"). "6-out" means that you get six out of a "gill" or "quartern".

As there are $5\frac{1}{3}$ gills to the bottle—six times $5\frac{1}{3} = 32$, so:
There are 32—6-out measures to a bottle; or,
There are 16—3-out (doubles) to a bottle.
Wines are generally sold by the half-gill; thus there are $10\frac{2}{3}$ measures to a bottle of Wine.

PRICE LISTS

From 1 December 1975, any intoxicating or other liquor offered or exposed for sale (i.e. visible to customers) in a bar for consumption on the premises must have its price displayed. This requirement includes drinks sold on draught if the dispense points can be seen by prospective purchasers.

It will be the licensee's duty to see that prices are displayed, which may be by way of a list, marking on optics, bottles, pump handles, shelves, etc.—or combinations of any of these methods. Barmen will have to be careful to see that when they replace bottles with price labels, the new bottle is labelled. If shelves are price marked, they must see that bottles are not in the wrong place. They must also be sure to charge customers with the advertised price, otherwise they could be prosecuted. Of course, errors in addition do happen with inexperienced barmen. If a barman is challenged by a customer when he gives him his change, he should quickly re-price the "round" and, if he has made a mistake, apologise and put the matter right.

BOTTLES

Wines, especially those which do not enjoy a quick sale, and particularly French White, Hocks, Clarets and Burgundies, should always be laid down (i.e. on their sides) and kept that way. This ensures that the liquid is covering the face of the cork and keeping it swollen. Failure to do this results in the cork drying out and shrinking, probably allowing the air to get to the wine and sending it "off". Air is the deadly enemy of all wine.

When your bottles *are* laid down, keep an eye on them for "weepers". These are bottles insecurely corked. If you touch them just under the neck and discover any are wet or damp, then the liquid is seeping out—and, correspondingly, the air is getting in! They should be removed from the bin for immediate sale, or re-corked.

Bottles when laid down should always have the labels uppermost. Not for fun, but for a specific purpose. Wine (especially Table Wine) will throw a slight deposit if left for any length of time. This deposit may take the form of crystal-like particles, flakes, or a muddy deposit, which will settle on the underside of the bottle, and is better seen through the clear glass, than through the label.

Now a few words about the "pushed in" bottom of some wine and spirit bottles, i.e. Champagne, Burgundy, French White, Claret and the Brandies. This is known as a "Punt" a "Push Punt" or the "Punt End".

There is a popular theory that this is just another way of "short changing" the customer on the quantity in the bottle. This is not so. Many years ago, wines and spirits suffered considerably more than they do today from the effects of the "finings" used to clear them and consequently wines very often required to be strained through a wire gauze strainer, or muslin, at the time of decanting. The "push punt" was there to collect the sediment and hold it back whilst pouring was in progress.

In very early Wine and Brandy bottles the "punt" was quite a large affair, extending well up into the interior of the bottle likes a mushroom. Gradually the shape of the bottles altered, the "punt" became smaller, until today it might almost be called "traditional"—as it is only in the last twenty-five to thirty years that "finings" and filtering have been improved to the point of performing the job satisfactorily.

There is no reason, other than tradition, for bottling Hock in tall amber-coloured glass bottles, and Moselle in the same shape made of green glass.

General Notes

Spirits in bottles are always stored standing up to prevent the spirit eating into the cork, but the cork itself may dry out in consequence. With the patent "Cork-in-Seal" spring-operated cap (as used by Martell on their three star Brandy) this is obviated—but still the bottles *MUST* be kept standing up.

One further point regarding Brandy old in bottle. Sometimes the corks become iron hard and most obstinate to move. An ordinary corkscrew is dangerous to use in these cases, due to the risk of breaking the neck of the bottle—even if it *is* being held in a cloth. Use the double-handled, lever-action corkscrew mentioned on page 77.

In the event of any bottle (Wine, Spirit or Beer) breaking at the neck or shoulder, decant the contents into another similar bottle (clean of course!) through a wine or spirit funnel. These have a fine gauze mesh to hold back everything except the liquid.

Do not be too concerned even if you have no funnel, for any chips of glass in the bottle drop immediately to the bottom and will *NOT* remain in suspension in the liquid—in spite of comments from uninformed customers about not wanting their mouths cut to pieces. Glass is almost as heavy as marble! and the finest sliver goes straight to the bottom immediately and does *NOT* float about. So, if no funnel is handy, pour the contents very slowly into a jug (clean) and thence into another bottle, but leave just the last teaspoonful in the damaged bottle. For your own peace of mind, and if you are in any doubt about this, just try getting a small piece of glass out of liquid in a bottle into another bottle pouring slowly. It is harder than you think—unless you turn the bottle right upside down to start with.

PART III
LIQUEURS, APERITIFS AND BAR ANCILLARY ITEMS

11
LIQUEURS AND CORDIALS BITTERS AND APERITIFS

There will be on sale in your house a variety of liqueurs. These are expensive and must be dispensed very carefully in the correct size glasses. An oversize Liqueur so far from showing any profit to the proprietor may result in a loss if sold in the wrong size glass, or at the wrong price.

Originally produced in the Monasteries by the Brothers of various Orders, they were made and dispensed to the sick poor to stimulate the circulation, and warm the system, more as medicines than for any social purpose. Several still retain the name of the Convent or Monastery, as in the case of Benedictine and Chartreuse.

Herbs, seeds, roots, etc., were crushed and steeped in water or spirit. Those made with water were called "simple" and those made with spirit as "strong water".

The word "Liqueur" may have been derived from "Elizir" or "Elizir Vitae"—meaning the "Water of Life".

Brandy is largely used in the making of liqueurs, but grain spirit or other alcohol finds its way into some.

There are, in some cases, a great many varieties of one particular kind of liqueur. Curacao, for instance, is produced by many distillers each with their own methods and ingredients. Because a bottle is labelled Curacao there is no guarantee that a Curacao by another firm will closely resemble it. Cherries, from which many Liqueurs are made, also vary greatly.

Cordials include:

Sloe Gin	Orange Gin
Damson Gin	Ginger Brandy, etc.
Lemon Gin	

and the difference between a liqueur and a cordial lies in the fact that a liqueur is a distillation of various substances designed to give flavour or perfume, whilst a cordial is obtained by steeping or mixing aromatic ingredients with spirits.

The word Cordial is often used in connection with *non-alcoholic* drinks, but this is misleading, these items generally being made from fruit compounds and syrups.

Again, a learned treatise on the many kinds of liqueurs could never be complete, because the recipes for most of them are jealously guarded secrets, perhaps known only by one or two persons, but the following will show the predominant flavour or ingredients of those most likely to be found on the shelves of the public-house. This is important, because quite frequently lady customers particularly will ask —"What is in that funny-looking bottle, and what does it taste like?"

THE MAIN LIQUEURS

Liqueur	Flavour
Advocaat (Pronounced ADVOCAH)	
This requires to be shaken vigorously before serving to mix the ingredients.	Brandy and Egg Yolk.
Apricot Brandy	
Made from ripe apricots with the addition of Brandy.	Brandy and Apricot.
Benedictine	
Sometimes incorrectly called D.O.M. which means "Deo Optimo Maximo" translated as "To God, the best and greatest". It appears on every bottle and is the motto of the Benedictine monks who make it.	Angelica and Herbs.
Chartreuse, Green (pronounced SHAR-TRURZE).	
Made by the Carthusian monks, this is considered to be the finest of all Liqueurs (after very old Cognac).	Angelica and Herbs.
Chartreuse, Yellow	
Inclined to be a little more "fiery" than the Green variety.	Angelica and Herbs.
[*Chartreuse, White* Now non-existent.]	
Cherry Brandy	
The British product made by Grants	Cherry.

of Maidstone and known as "Queens"
is alcoholically stronger than most of
the imported ones.

Cointreau (pronounced KWAN-TRO)
Colourless and enticing.   Very dry
                                                  Orange.

Crème de Cacao
(Pronounced CRAME DE CACKO)   Chocolate.

Crème de Menthe
(Pronounced CRAME DE MONT)
Brilliant green—a great aid to the di-   Peppermint.
gestion. This liqueur will lose colour
and fade if kept in bright sunlight.

Curacao
(Pronounced Cure-a-so).   Orange.

Drambuie
One of the most popular of all Liqueurs.   Whisky and
Made in Scotland.   Angelica.

Grand Marnier
(Pronounced GRA-MARNIER)   Orange.

Kummel (pronounced KIMMEL)
A great favourite after dinner, it is an   Carraway
aid to digestion.   Seed.

Maraschino (pronounced MARAS-
KEENO)
Dressed attractively in square straw-   Almonds.
covered bottles it is distilled from a
bitter cherry grown in Dalmatia (a
coastal district of Jugoslavia) but is
also made in this country. It is white
and very sweet.

Peach Brandy
Obtained from the peach which is   Brandy and
steeped in Brandy.   Peach.

Tia Maria
An irresistibly seductive Liqueur. Given   Coffee and
the opportunity the ladies will drink   Rum.
half-a-bottle.

Van der Hum   Orange.
From South Africa.

Of all the Liqueurs Advocaat is the cheapest and should
be served in wide-top glasses. As stated, the bottle should be
shaken vigorously before serving, the Brandy being inclined
to separate and be found at the top. If these bottles are

turned upside down overnight when apparently empty half-a-measure can be found in them in the morning.

Two drinks may be called for which include Advocaat— "Angel's Kiss" and "Snowball".

"Angel's Kiss" is a measure of Advocaat in first and a measure of Cherry Brandy added. When mixed, "luvverly" though it is supposed to be, it looks like a glass of cocoa.

"Snowball" has slightly more to recommend it, being Advocaat diluted with Lemonade. It is dealt with under "mixers".

Advocaat varies considerably in colour, some being quite pale, but nevertheless quite as palatable. Warninks is a rich-looking one. It makes a wonderful sauce when poured over mince pies and Christmas pudding.

Next to Advocaat in popularity comes Cherry Brandy.

Green Chartreuse is the strongest and most expensive. At one time it used to be admitted at 5° *over* proof but now it is reduced to 96° proof.

Strictly speaking Liqueurs are not public-house drinks— they are all the better for being drunk after a meal, for which purpose they are admirably suited—except Advocaat which is a meal in itself! But human nature being what it is and money being free, do not be surprised if a customer chooses to make a beast of himself by getting drunk on Drambuie in your bar.

When serving Liqueurs always turn the bottle
back to front.

Any drips will then run down the back of the bottle and not over the label. As some of these bottles remain on the shelves quite a long time, it is nicer if the labels are not too soiled.

SYRUPS

There may be on your shelves strange bottles with stranger names which will probably be Syrups and Cordials. These are used in making up drinks—cocktails, etc.—and the labels will give you no clue as to what is inside. The following list gives the flavouring of those most frequently found and will help you identify the contents, though it is doubtful if you will find them in general use in the ordinary public-house although the names themselves often appear on Liqueur labels as "Crème de ...".

Fraise (strawberry)
Framboise (raspberry)
Orgeat (almond)
Cassis (blackcurrant)
Gomme (sugar)
Cacao (chocolate—not known as a syrup)

Noyeau (bitter almonds)
Menthe (mint)
Banane (banana)
Cerise (cherry)
Citron (lemon)
Grenadine (pomegranate)
Ananas (pineapple)

BITTERS AND APERITIFS

It is strange that hardly any Bitters give any directions whatever for mixing. The bottles are devoid of recipes or instructions and in many cases Bar staff have been known to dust them for years without knowing what they were or for what purpose they are used.

In the main Bitters are used as an ingredient in cocktails and mixed drinks generally—they are hardly ever consumed "neat", being diluted, even if only with soda-water.

Aperitifs, on the other hand, usually have some directions attached or on the label so that Bar staff are not left in too much doubt about their uses, composition and method of mixing. The term aperitif means appetizer, so pale sherries and most vermouths (dealt with under their own headings) can be called aperitifs.

Angostura Bitters

To give them their proper name—Dr. Siegerts Angostura Aromatic Bitters—come from Trinidad and contain many herbal and root extracts, the recipe for which (as in every similar case) is a closely guarded secret. It is *not* consumed by itself. It is the "pink" in Pink Gin (as explained above, p. 74) and it may be used to tint "Babycham" to a delicate hue besides giving it a pleasant bite.

It is well known as an antidote to a bout of over-indulgence and has succeeded in bringing round many a person who not long before was likely to be well "under the weather". The recipe for this, in case you are called upon to dispense it, is a couple of dozen shakes of the bitters (not more than half a tea-spoonful) in a spirit glass topped up with soda-water. It is not too pleasant to take and often results in the victim being sick and sobering up in double quick time. A well-known Radio star who was hopelessly drunk and quite incapable of making an appearance in front of the microphone was given this sovereign remedy and miraculously performed his act only thirty minutes later without a soul being the wiser.

These Bitters have a shaker top which fits all sizes but are sometimes decanted into small glass bottles (with a shaker top) and you should make enquiries about any such bottles you see on your shelves.

Underberg Bitters

This Bitters is a product of Western Germany and is famous all over the Continent as a "livener". It arrives here at 86° Proof and is sold in miniature bottles without any directions as to use being shown.

For those suffering from the evil effects of the night before one of these miniature bottles in a spirit glass topped up with soda-water has quite a remarkable effect in dispelling the head-ache, shakes and revolving stomach so much part of over-indulgence.

Fernet-Branca

On the market for over 100 years, Fernet-Branca is well known. It is recommended for use in three ways: added to coffee—a little in the coffee itself, as a digestive after a heavy meal or as an antidote to depression—a small wine glass with equal parts of water, or soda-water. It is a great pick-me-up and so far from containing anything injurious it is universally known that its ingredients have a very beneficial effect.

Orange Bitters, Peach Bitters, Campari Bitters

The above are all used in various drinks to give flavour and piquancy. Laws are famous for their Peach Bitters and Campari from Italy is famous everywhere. Served on its own, with vodka or gin, ice and soda, or with a spirit *and* soda, it is increasingly popular as an Aperitif. The following are among the better known Aperitifs:

Dubonnet

Forceful advertising has brought Dubonnet to the fore. It may be served in a variety of ways—several being given on the back label of each bottle. When consumed on its own (as it often is) it is improved by the addition of a slice of lemon. Gin and Dubonnet is quite a favourite—using about the same measure of the Aperitif as Gin—with a slice of lemon if requested. Dubonnet is traditionally red but Dubonnet *Blonde* (white) is now also on the market.

Pernod 45 (pronounced PAIRNO)

This drink breathes France. "What would Paris be like without Pernod?" ask many writers.

Many "old-timers" in your bar will recall Pernod but will be quite mistaken if they are confusing it with the Pernod 45 of today—for they will be thinking of Absinthe (pronounced AB SANT).

Absinthe is a spirit combined with extract of wormwood and used to be produced by Pernod Fils and imported into this country at a very high strength, around 100° Proof. Its manufacture was banned by the French Government because over-indulgence by some misguided people led to it being shockingly abused, which, of course, was not the fault of Pernod Fils of Pontarlier.

The Pernod 45 of today contains no wormwood but has Anise for its base. The herb Anise has been recognised for over a thousand years for its outstanding health-giving properties. Charlemagne, who became Emperor of the Romans and lived A.D. 742-814, a most wise and brilliant ruler, decreed that the herb Anise be planted in all his territories so great was the esteem in which it was held.

For a time Pernod was known as Pernod de Anise, but its new title Pernod 45 is now well advertised. It is imported at 78° Proof and is certain to be on your shelves. Don't forget to enquire the price at which it should be sold.

Years ago customers used to amuse themselves by placing a piece of lump sugar on a spoon, allowing water to drip over it and as this simple syrup spilled into the glass the Pernod would become opalescent. The same effect may be obtained today with Pernod 45. The sugar water has the effect of releasing the "ethers" in the Pernod.

The accepted method of serving Pernod is to dilute it in the proportion of one volume of Pernod to four or five volumes of iced water. Of course, if ice cubes are used then slightly less water is required as this makes for additional dilution.

NOTE: Pernod is also excellent with Bitter Lemon and is often ordered with lemonade.

Amer Picon

This is an Aperitif unchanged since it was first compounded in 1837. It is pleasant and most refreshing, con-

111

taining Orange, Gentian, Quinquina and Sugar.

As a long drink Amer Picon is improved by being decorated with fruit and the recommended method is to use the attractive "devil's trident" which is supplied to spear alternatively cherries, a slice of orange and a slice of lemon. Served in a large goblet with a couple of ice cubes and a "double" of Amer Picon topped up with sparkling orange it looks most attractive—and tastes even better. For the lady who can't make up her mind what to drink you should recommend it as something "different". For the lady who has already tried it and found it enjoyable you will be able to mix it and not be caught out when she (with that idea in her mind, no doubt!) requests you to make her a "Picon-Picon"—by which name it is known when served as a long drink.

Amer Picon is also much used as a mixer, i.e. served with any of the soft drink mixers and as a base for cocktails and other short drinks.

St. Raphael

This wine aperitif which is no stranger to the visitor to France with its attractively shaped bottle of the original design is commonly found in our bars today. There are two kinds—the well-established White and the newer Red which is not too dry, not too sweet, nor too bitter. St. Raphael, based on a very old recipe, has remarkable tonic properties.

In 1830 a blind Frenchman named Jupet perfected an aperitif with a cinchona base (cinchona or quinquina is a bitter-tasting bark of Peruvian origin). The story goes that he prayed to Saint Raphael, the patron of the blind, to restore his sight, promising that if his prayers were answered he would name his aperitif in the saint's honour. His sight was restored and his recipe became known as St. Raphael. Jupet's recipe was in fact based on an ancient custom of taking the powdered bark of quinquina as a tonic.

St. Raphael may be enjoyed at any time of the day or night by itself with ice and a piece of lemon. As a long drink add ice, soda-water, tonic water, or bitter lemon. If it is required as a stronger drink add vodka, or gin.

12
CIGARS, CIGARETTES, SNACKS

CIGARS

These will be on sale at various prices—from Whiffs to Presidents. Many are now packed individually in metal tubes and cellophane, which serve two useful purposes. The save the cigar from damage. They prevent it drying out.

It is not likely that you will be selling fine Havanas— probably Jamaican, Dutch and certainly British rolled cigars —the biggest seller being the now popular Panatella—a long thin shape selling reasonably cheap.

If you have cigars loose and unwrapped in a box it is usual to present the box to the customer and allow him to make his selection. If they are individually wrapped the cigar may be handed to him.

Corona, Petit Corona, Panatella, Torpedo, etc., are all shapes of cigars *not* the cigars themselves. The strengths are marked on some boxes using the letter "C", thus C (Claro) is quite mild; CC (Colorado Claro) is medium strong and CCC (Machero) is full strength.

NOTE: A cigar should always be *cut* to allow a free draught. This cannot be properly done by piercing, as the smoke and tarry substances become trapped at the tiny hole, moisture collects there, and the cigar becomes intolerable after a short time.

The best method is to cut out a "V" shaped wedge almost the whole width of the cigar, either with a proper cigar cutter, or with a very sharp knife—a razor blade will do. This deep cut will make even a cheap cigar smoke sweetly.

Cigars should never be dry—or allowed to get dry. Therefore they should be kept in the cool. The silly habit of Betting Shop barons of dipping them in their beer to dampen them and prolong the agony is a filthy and dis-gusting insult to the beer!

A cigar should always be lit with a match—evenly, all round the edge—and smoked slowly.

CIGARETTES

Make sure you know what brands you stock and their respective prices. If you are *repeatedly* asked for a brand unstocked, make a note to tell the Guv'nor; he may decide to stock it.

He won't, of course, be able to carry anything like enough brands to satisfy all tastes, so you must learn what to offer a customer when you haven't got his favourite. To reply to a man who asks for Silk Cut, "No, sir, but we've got Capstan Full Strength", would be as daft as offering a glass of port to a man who asked for a medium sherry. Get the brands firmly fixed in your mind according to *groups* (including the more popular brands you don't stock but are likely to be asked for) according to (a) price; (b) Tipped or non-Tipped; (c) Mild or Strong.

When you are setting out cigarettes, *do* put them *all* up the right way. This may seem niggling—but just think what it would look like if you went into a Supermarket and saw all the packets upside down on the racks. Separate the cigarettes as you stack them. Sometimes they are inclined to stick together, especially Senior Service, and as one packet is removed a whole load comes with it—like a string of sausages.

A practice which has become common in some bars concerns the protective wrapping on cigarettes.

It is not correct to remove wrappings without asking

for three very good reasons:
1. The cigarettes may have been purchased for some other person.
2. They may be accidentally dropped into a puddle of beer when being handed over.
3. You may have made a mistake over the brand required or the customer may change his mind, *after* you have stripped it off.

Some barmen go even further and, after stripping off the wrapping, open the packet and remove the silver foil as well. This practice started in hotel lounges and exclusive clubs by waiters offering a "service" to the customer in rather the same way as they would offer a light. It has now become common in many pubs, but many customers resent it.

It is true to say that the cellophane wrapping once off the packet, is a menace. That taken from a packet of twenty will completely fill an ordinary ashtray. It will, if dropped into the "gents", block the channel. It is no decoration for a carpet if dropped on the floor.

The best place to dispose of it is your side of the bar but do enquire first whether you *may* remove it.

Children and Cigarettes

No child under sixteen years of age may be served with cigarettes or tobacco. It is not often that a child of a tender age will be found rolling cigarettes so it must be fairly obvious that the ounce of shag she comes in for is for some other person—but there is no knowing how old that "other person" is. With the present drive against children smoking you will be doing them a favour and respecting the law if you do not serve anyone apparently under sixteen with cigarettes even in your off-licence department.

SERVING SNACKS AND ANCILLARY ITEMS

Whilst it is an established fact that any item containing fat or grease tends to turn beer flat it is an equally established fact that your customers will demand snacks—flat beer or no flat beer!

Probably the foremost among these is the bag of crisps in one or more of their various flavours and you will doubtless be astonished at the number of customers who ask for Smith's and resolutely refuse to be tempted by any other make (however good), even to the point of "going without".

Remembering that the contents weigh less than one ounce it is truly surprising how satisfying they are. Very few people manage to get through two bags—unless they haven't eaten since the day before or are candidates in a crisp-eating contest.

Their popularity remains unimpaired over the years, being, as they are, just sufficient to "soak up the beer", and they will be found on sale in almost every public-house. Being in sealed packets they do not carry any risk of contamination from germs, etc., as perhaps might be the case with a sandwich or sausage which is not wrapped.

It stands to the credit of the manufacturers of crisps that there has never been a single case before the Public Health Authorities.

It will be your duty behind the bar to ensure that all pre-packed foods are sold in strict rotation, and that crisps, in particular, are kept in their polythene bags. A crisp that isn't crisp isn't a crisp. It will probably be your job to see that you are stocked up with these items so make sure you do so *before* your bar opens.

You may also be dealing with biscuits, nuts and raisins.

If you are called upon to serve a sandwich do make sure to use tongs when handling it. Never, on any account, use fingers for any foodstuff—it is highly insanitary and dangerous. The risk of contamination is ever-present. Wash your hands every time you have handled anything of a dirty nature and particularly after the toilet.

It is important at *any* time to have spotlessly clean hands and finger nails—when you are handling foodstuff it is *DOUBLY IMPORTANT*.

**Repeat again—after using the toilet
wash your hands.**

PART IV
GENERAL BAR PRACTICE

13
ON THE JOB

STARTING WORK

It is perhaps as well to emphasise right at the outset that the Licensed Trade is a *domestic business* and is not like any other trade.

Centuries ago (and in villages even today) the tavern was a private house, with just one room open to the public—very often the kitchen.

Beer was brewed on the premises by the woman of the house who was known as an "alewife" and a pole was exhibited to show that it was an alehouse or tavern rather as some barbers' poles are seen today.

In Roman times there were "tabernae" along the roads. If the "alehouse" sold wine as well as ale a wreath of leaves was hung on the pole and this gave birth to the old saying—"Good wine needs no bush"—meaning there was no need to advertise good wine—the customers came to seek it. Thus we find the commencement of the inn sign, fast disappearing nowadays which is a pity, because many were pictorial and decorative besides giving the essential information—the name of the house and possibly the date. But today it is possible to walk round and round trying to find out the name of the pub.

Well, the public-house being domestic in nature can no more suffer hard and fast "trade union" rules than could an ordinary household.

Many jobs unconnected with the actual service of drink have to be attended to in exactly the same way as they do in the home. You do not allow the Lounge Bar fire to go out for want of a scuttle of coal, the Public-bar clock to stop for want of winding, or the Saloon-bar sink to remain stopped up without making some effort to clear it.

When you take up your duties whether you live in or out, part-time, or full-time, you must be prepared to undertake some menial chores—and not grumble about it.

In all houses the physical work (cleaning, etc.) is done in the morning and barmen should provide themselves with a pair of old slacks and shoes and barmaids an overall.

On taking up your job you should bring with you your National Insurance Card and P.45.

NOTE: (a) If you are working elsewhere and only part-time in the public-house the person, or firm, who employs you *first* during the week is responsible for your National Insurance—irrespective of your earnings at either job.

(b) With regard to Income Tax (P.A.Y.E.). If you have other employment and pay tax your part-time employer should inform the Tax Office about your employment with him. It often happens that part-time workers (if they are already paying tax) are given a Code Number[1] which means they will pay the full rate without any allowances in their new employment. Be warned! It is far better to pay any tax due on your part-time earnings week by week (even if you over-pay!) than be caught at the end of the year with all the arrears.

SPIT AND POLISH

The following items will be found very useful and you should always provide yourself with them when starting a new job. (They will probably be available on the premises, but it will save you standing about whilst they are dug out of some nook or cranny or can't be found at all, if you have them with you. And you will create an excellent impression.)

Two or three cleaning rags

A small piece of domestic yellow soap—unscented, *not* carbolic

A small piece of steel wool—*not* one impregnated with carbolic soap

A small quantity of brass polish

A piece of chalk

A scribbling pad and pencil.

Nothing creates a better impression than the person who steps smartly behind a bar for the first time and goes to work with the minimum of instruction.

---

[1] The Code Number is what tells the employer what he must deduct from your wages and pay over to the authorities on your behalf.

The "Pewter"

This is the shelf under the counter into which is inset a sink (known as a "well") on which glasses are stood to drain after they have been washed and then returned to their customary place after being polished.

This shelf is known as the "Pewter" but in modern houses where the shelving may be of stainless steel or Formica, it is sometimes called the "wash-up".

After a busy morning session the pewter will be found sticky from beer, etc., and will require a good wipe down, otherwise the glasses will stick to it. Wiping down is usually done as and when convenient during the session.

At the close of the evening session it is the common practice to wipe and dry the top of the counter, spread out sheets of newspaper, or glass cloths, and stand all the glasses thereon—upside down, of course! This gives a clear start in the morning, both for washing and shining the pewter and making sure that *all* glasses receive a polish.

The newspaper or glass cloths prevent any damp glasses from leaving a ring mark on the polished counter top.

Spirit glasses have been seen to burn into counters a ring which is there for good and cannot be removed by any amount of polishing or scouring.

NOTE: In some cases the above procedure cannot be observed—mainly where relief staff have to catch last transport home—but it is still bad practice to leave it till the morning.

Cleaning the Pewter

If it is real pewter it is no easy job and requires plenty of energy. It should gleam like silver, but if it has been neglected (and many of them are!) it will take a long time, weeks perhaps, before it is brought up to standard.

Probably the most effective method of cleaning pewter is to wash it first with hot soapy water, or an abrasive powder, to remove all sticky matter, rub it over with damp soapy steel wool (No. 0), dry off thoroughly and then rub it vigorously with newspaper to produce a shine. Some bar staff use proprietary metal polish, but unless this is completely washed off it has a tendency to taint the beer, as will carbolic, pine fluid or scented soap. Beer is very susceptible to any strong-smelling substance.

The Glasses

After the pewter has been shined the glasses require polishing before being replaced—and that *means* polishing! Nothing makes a barman look so stupid and incompetent as having a dirty glass or one smeared with lipstick returned by a customer.

Glasses are always stood upside down both on the pewter and on the shelves, however pretty they may look the right way up. The only exception to this rule is in the case of those on display for purely decorative purposes and even then there is always the risk that one will be picked up in a hurry and drink served in it when it is dusty.

After all, this is where you, the wine grower and the shipper get together. *You* are the very last link in the chain between production and service. All the loving care, all the experience of viticulture,[1] all the waiting and maturing, all the money and hard back-breaking toil, disappear up in the air if *you* serve wine in a dirty glass. You *must* do your share and see that as far as you are concerned *perfection* is the watch-word. The wine shipper will not let you down—his work is already perfect when he corks the bottle—the rest is in your hands.

Good wine is to be contemplated, considered, examined for colour against the light, sniffed for bouquet, rolled round the tongue and lovingly treated in every way. None of which can be accomplished if the glass in which it is served smells of a stale dish cloth—or is anything less than sparkling bright.

The Counter Top and Bar Tables

If the counter top is of wood or lino it will require polishing. The polish you apply doesn't do the work—you do! Formica tops are wiped and dried.

In nine cases out of ten the underneath edge is forgotten—both outside and inside the bar. Along this edge spilled beer drips and eventually congeals into a sticky tarry mess, therefore *it must be thoroughly wiped* on both sides of the bar. If it is left too long it may have to be chopped off!

Many dresses and suits have been ruined because customers, sitting on a bar stool, have caught a knee under the edge of the counter top and received a black mark on their clothing which rarely comes out in cleaning.

[1] Viticulture is the husbandry of the grape and the cultivation of the vine.

Similarly, if you are required to wipe the bar tables don't forget to do the edge—and *underneath the edge* as well. (Incidentally this is one of the things supervisors of company houses watch for.)

In some houses there is a shelf running along the wall for customers' glasses and this will require wiping at the same time—as well as the mantelshelf and the piano, upon whch some customers cannot resist resting their glasses—in spite of notices asking them to refrain from doing so.

Make sure all your ashtrays are clean—and that there are enough of them.

Bottling-up

You may be asked to assist in filling shelves with bottled beer. Remember a most important point which cannot be stressed too often:

Never place new stock in front of old.

It has been known for bottles to remain at the back of shelves for weeks and in the end become unsaleable. The shelf life of the light, carbonated beers (Pale Ale, Brown Ale, etc.) is very limited and the Stouts go sour. Lagers and "heavy" beers last longer—but even so it gives a bad impression to serve any beer from a bottle covered in dust.

Always dust each bottle as you bin it away. A quick wipe round the shoulder is enough to remove the loose dust. Before placing them on the shelf give *that* a good dust as well. In the dim interior of the house it will probably not show up so much, but let a shaft of sunlight strike the dusty shelf and bottle and the picture is one of neglect and sloth.

Point number three. Line up *every* bottle to face the front—not just the front row! Bottles placed on the shelves with military precision, clean and sparkling, look "on parade", compared to those just shovelled on anyhow and the extra time involved is negligible. Unless you line up the back ones as well nothing will be achieved because once the front line has gone they *are* the front line.

It is true that when your bar is packed with waiting customers absolute priority must be given to taking their money, straight bottles or not, but when you slacken off see they are straightened out properly. The foregoing also applies to those bottles kept under the counter: just because the customer can't see them is no reason to slack off with the chore —and it is to your advantage to be able to read all the labels.

Ullages

When "bottling-up" you will come across odd bottles unfit for service. Sometimes they are corked but empty, having missed the bottling machine, others may be cracked or damaged. Do not just leave them in the case. They should be given in to the cellerman for return as "Ullage" for which an allowance will be made by the brewery. (Some breweries give an over-all allowance off the invoice to cover all ullages irrespective of whether there are any or not.)

The Cabinet and Display

There is a curious idea prevalent with some bar-staff that the moment the bars are open for service all cleaning stops.

It is true that the cleaning of floors, brass work, pewter and shelves should never be in evidence after opening time —but very often it is physically impossible to arrange for the cleaning of display cabinets to be done except during "hours". Before commencing, however, you should enquire if the morning trade is slack enough to allow for this to be done. If it is done during hours do not place full stock on your bar counter—keep it well out of reach of customers.

Beware of the loudmouth who, when he sees you working, taunts "Guvnor's man" and "You'll win no medals for doing that." He won't get so far himself in the end. Ignore him.

There is only one way to clean and polish a bottle to ensure a real shine and that is with a damp chamois leather and a dry clean duster. Bottles "on ullage", i.e. your Ports, Sherries, Liqueurs, Squashes and Cordials, etc., get very sticky and require to be well wiped. The lazy barman with a dirty piece of rag would probably do much better to leave them alone.

Never damp the labels—avoid them—unless they are really bad or happen to have a varnished finish—and even then be very careful. Some liqueur bottles remain on the cabinet for months and even years and in a very short time the labels may become unreadable if they have been constantly wiped over with a wet cloth. This applies especially to neck labels and bands.

Don't forget to wipe and clean the "push punt" or "punt-end" of all bottles made that way. This is the identation in the bottom of some bottles—Burgundy, Champagne, Brandy, etc.

Your glass shelves and mirror back may be polished with Windolene, but watch for the fine powder dust as you wipe it off, because it is inclined to settle on your polished bottles.

It is a good idea to take your duster outside and shake it periodically and to polish your bottles *after* you have done the shelves and mirror back.

Glass shelves have *edges* and these should not be forgotten. If your shelves are removable *take them down* for cleaning. Do the job properly.

NOTE: To keep a chamois leather from going hard and breaking up leave slightly damp in a jar or tumbler and do not let it get dry.

Wine and Spirit Measures and Metal Tankards

If these are shined with metal polish wash them afterwards in hot water and then dry them. This will not affect the shine but will rid them of the taint of polish. Make absolutely certain of this by smelling them carefully before they are replaced for service. It is better *not* to clean the inside of measures or tankards with metal polish—or any strong-smelling agent. Shine them up with a damp swab—unless, of course, they are in a very bad condition when metal polish may be necessary—but wash them well in hot water afterwards.

Spirit Measuring Taps

These are often wrongly described as "Optics". There is only one optic—the "Optic Pearl" patented by Gaskell's which is a measuring tap with a heavy glass bull's-eye which may be seen emptying and re-filling with each forward or backward movement of the lever. These "Optics" dispense the correct measure each time the lever is moved and completely superseded the thimble measures previously in common use. Later came the "Non-Drip", a measuring tap with a one-hand action. The spirit glass is merely pressed on to a bar which releases a measure at a time and two adjacent "Non-Drips" can be used at once as both hands are free to hold the glasses. With the "Optic Pearl" one hand holds the glass while the other works the lever.

NOTE: When putting up a new bottle of spirits with a "Non-Drip" tap make absolutely certain that the spring-loaded catch has secured the bottle to the stand. Give the bottle a forward pull holding it firmly whilst doing so, to ensure that it is securely fitted in the holder, otherwise it may slide out during service.

It is as well not to attempt cleaning the interior of the "Optic Pearl" before you know something about it—and certainly not unless you have the necessary rubber washers available. It is possible, with the proper tool, to unscrew the front of the "Optic Pearl" and clean the back of the glass and the face, but the glass is sealed with a rubber washer which invariably requires renewal. It is most important not to cross-thread the front when replacing the glass.

It is *not* wise to attempt cleaning the interior of the "Non-Drip" Tap; these can be properly serviced by Gaskell and Chambers.

The "Bowker" Pourer and the "Dalex" may be dismantled quite easily and the bowl and interior cleaned. *Do not* use any metal polish or steel wool on the interior of either of these pourers. (See Glossary of Terms for description of these pourers.)

The interior of those used for rum will turn quite black as will the others if they are not cleaned regularly. Once a week is the minimum to avoid corrosion of the metal—but better still—leave them in water each night, but be certain they are quite empty first. The "Bowker" often takes a few up and down turns before it is completely empty so:

Always make certain you have completely drained your pourers of spirits (back in the bottle, of course!) before you start to clean them.

and

Likewise make absolutely sure all water has been drained out of them before they go back in the bottle. Sometimes they appear to be empty but are still holding some liquid. Unless they are absolutely clear of water your first customer may receive a very diluted drink (bad enough!) but he might be a Customs and Excise Officer (that's worse!).

Swabs and Glass Cloths

Always keep your swabs clean and white. Wash them out regularly and leave them to soak in whatever bleach you have on the premises—after which rinse them thoroughly. Make sure your glass cloths are clean and keep them clean. Hang them up when not in use to give them a chance to dry out. Keep all cleaning materials out of sight.

THE SERVICE

When you are first delegated to your post behind the bar your piece of chalk will come in handy for marking up the prices. As these soon get washed off make a note of them on your scribbling pad for future reference. Make sure of all your prices, wines and spirits, spirits with squashes and especially shandies and the size generally used.

After you are certain of your prices take note where the different kinds of glasses and measures are kept, and such things as Lime Juice, Orange Squash, Ports and Sherries, in fact be nosy! Enquire about ice, lemon, a knife and tongs, cocktail sticks.

Make sure the bar is opened on time. It often happens that a bar door is left bolted and forgotten until someone bangs on it (usually with some rude comment).

The cellarman should have "pulled up" the draught beer ready for service and now you are waiting for the "off".

It is assumed that you look smart, clean hands, clean nails, clean collar, clean jacket (or overall).

Make yourself familiar with the Till.

As your first customer approaches greet him with a

cheerful "Good morning, sir!" not like some who just grunt "Huh?", others who say nothing and wait for the customer to speak, or still others who don't even look up from a newspaper and wait for the customer to bang!

Remember this: even as a Barman you are the host and the customers are your guests. By opening the bar doors you have invited them to come in. It would be a sorry thing if you were invited out to someone's house for a drink and your host didn't open his mouth to greet you, or just stood glaring morosely at you, or worse still just went on reading a newspaper. You probably wouldn't accept a second invitation.

Never keep a customer waiting—it is most annoying. It will not escape you that a man quite resigned to wait ten minutes in the Post-office for a stamp will shout the place down if he is kept waiting more than five seconds for a drink. In fact many seem to think there should be one bartender to each customer!

However, should you be engaged in some job more important than taking his money (if there *is* anything more important!) always acknowledge him and say you won't keep him a moment. By doing that he at least knows you are aware of his presence. It is not unknown for a clever barmaid to keep up a running commentary with half-a-dozen customers all of whom are quite happy to await her service and, in fact, even seem to enjoy it!

THE PUBLIC BAR

Some staff enjoy serving in the Public Bar better than the Saloon or Lounge. They appreciate the "earthy" touch of the "honest-to-goodness" working man, the quick and snappy conversation, the everlasting "mickey taking".

Should it be part of your duty to serve in the Public Bar you may have to suffer a certain amount of ribald comment from the regulars.

" 'Ow long *you* gonna stay? 'Ad eight noo barmen 'ere in six weeks! "

Probably untrue anyway. The "Public" as they are called are very fond of "having a go" at anyone new but you should just laugh it off and not get bad-tempered. When they fail to get a rise out of you they'll go back to their dominoes.

They will, however, watch you like a cat with a mouse, hoping to catch you out, mainly because they have nothing better to do.

Wrong change rates three roars all round the bar—but *only* if it is short. A mistake in service, i.e. pulling up Mild instead of Bitter, rates two roars. Short measure is good for five minutes' hollering and hooting.

If you are bothered because you cannot understand an order given in slang, or Cockney rhyming slang, ask the customer nicely if he would mind giving it in plain English. Some London types think it a huge joke to pick on a new barman with such an order as:

"Pint o' Diesel, an apple fritter and a Tom Thumb."

You would be quite in order to ask the customer to translate it for you, when you will find it means, "A pint of Mild Ale, a small bitter and single rum."

Ale is known variously as : Mild, Ale, Double XX, Diesel, Splosh, Hogwash, and in some districts simply as "Beer".

Bitter is sometimes called Apple Fritter.

Scotch Whisky may be called Pimple or Pimple and Blotch, Hooch, Gold Watch.

Gin will be called Needle or Needle and Pin, Vera or Vera Lynn, Mother's Ruin.

Rum will be called Tom Thumb, Nelson's Blood, Black Jack.

Brandy will be called Coconut Candy.

These are London terms—various parts of the country

have their own slang names for drinks.

In your Saloon or Lounge, of course, you meet rather a different type though the working man often favours the Saloon Bar in these days. So now you are taking your first order—a round of drinks and other things.

Before serving anything look quickly at each glass to make certain it is not cracked or chipped and that it is sparkling bright. Before serving wines or spirits give each glass a quick polish with your glass cloth. A little bit of flourish, a little bit of "madam", a little bit of showmanship goes down better than anything furtive and "under the counter". The little bit of polish does not go unnoticed—especially by the customers who matter—and they *all* matter.

THE ORDER

Always, repeat *always* add up your order as you go along.

Add one item to the other. On no account draw up the drinks and *then* attempt to reckon up what it came to. At busy times drinks may be taken off the counter and handed round unseen and not added into the bill. Customers themselves are not very helpful (and regrettably, some are downright dishonest). You may be given an order in bits and pieces specially designed to trip you up. Here is a specimen order given over the bar of a London pub recently. It is a difficult one—even for an alert bartender.

"Pint of **Best Bitter**"

"Brown Ale"

"Gin and Bitter Lemon"

"Another pint of Best Bitter"

"Two Pale Ales"

"Scotch"

"Twenty Players"

"May I have the Soda Water?"

"Guinness"

"Will you change this Brown Ale for a Ginger Ale?"

"Bag of crisps"

"Box of matches"

"Will you change these crisps for cheese and onion?"

"How much?"

Should you leave the adding up of this little lot until the whole order is completed you will probably need a pencil and paper—and an accountant! . . . and then the chances are you will be wrong. Half the drinks may have been taken by this time to the other end of a crowded room (or even consumed) so you won't be able to check.

Adding the order up as you go along makes it comparatively easy—although it is no easy one, not even for a very experienced barman.

Only by adding up the order in this way are you able to tell the customer the amount he owes at any time.

Remember it is the *only* way to do it.

Now there are three traps in this order. Have you spotted them? They may be there by design or just accidental, but in either case they may well put you off adding up. If by design the customer is hoping to whip one item off the counter (or perhaps the lot!) and leave you floundering with your arithmetic, his chance comes three times. First when he asks for the soda-water. Second when he changes the Brown Ale for a Ginger Ale and third when he asks to change the crisps. These little diversions tend to muddle up the counting.

Always be particularly alert as soon as you realise you are facing a large order, because the customer may query the amount charged and ask you how you arrived at it. It is a good idea for you as a beginner to ask a senior member of the staff to check over any doubtful order, or, if time permits, jot the items down on a pad. There are some customers who delight to query each and every order and often if they find it correct they will then query the change. A good licensee will have a price list displayed so that everyone can check.

The Change

Which brings up to another point: "ALWAYS "ADD ON" THE CHANGE. This may appear elementary to you but it is surprising to find so many learners who don't do it—or take a long time to understand it.

Your order comes to, say, 77½p and the customer has given you a one-pound note:

ADD ON ½p — making	78p
ADD ON 2p — making	80p
ADD ON 20p — making	£1.00

Beginners often start by trying to work out the amount of the change subtracting 77½p from one pound. Having succeeded in arriving at 22½p as the change, they have only to have their attention diverted when they will forget which amount represents the order (77½p) and which the change (22½p) because they are dealing with *two* sets of figures. By *Adding On* the change to the amount of the order as shown above the actual amount of the change given is probably not known and it doesn't matter. All that matters is that the customer has received a sum of money which, when added on to the amount to be taken, comes to one pound.

Always "add on" the change to the amount of the order.

When taking orders and giving change avoid all superfluous chat. It is most distracting and is sometimes fostered by customers with that idea in mind. The fact that Carol is having a baby may be welcome news or add up to a juicy bit of scandal but it won't help you to add up an order—or give the right change!

SPECIAL NOTE: There is a special point to watch in the case of mixed drinks when both are "draught", i.e. mild and bitter, etc.

For instance: The customer requires a pint of Mild and Bitter. The Mild Ale is 18p per pint; the Bitter is 24p per pint. The correct and easiest method is to add the two amounts together and then halve them, thus:

18p and 24p equals 42p. Half 42p equals 21p and 21p is the price you will charge. Unless you learn to do it this way you will finish up looking very dim if the customer should only want *half-a-pint* of Mild and Bitter. All you have to do then is to halve the price per pint, thus:

Half 21p equals 10½p and 10½p is the price you will charge.

SILLY NOTE: Do not hand change back right over a glass of beer. Many, many times it has been known for it to finish up in the bottom of the glass.

The Till

The manufacturers of Manual Cash Registers advise you not to be too delicate in ringing up. The keys have a fair amount of work to do inside the register and for that reason they should be struck firmly and sharply, for any half-hearted tap often causes a jam.

Should you make an error in ringing up, i.e. ringing 80p instead of 8p through touching adjoining keys, call the person in charge and point out the error immediately whilst apologising to the customer for the error and the delay. DO NOT TRY TO CORRECT THE ERROR without advising someone first and if you hold on to the customer's change then you have him as a witness to the error. Some bar staff "off-ring" subsequent orders until the amount of the "over-ring" is corrected. This is completely wrong unless it is done "on orders". It looks very fishy if you complete an order and the customer then reports to the Licensee that you rang up "No Sale"—however innocent you may be. The tills are there to protect you, the customer, the Licensee and, of course, the takings.

You till will contain a "Float". This is the sum of money you open with for the purpose of giving change. At the end of the session the "Float" is extracted, the till reading is taken and the cash in the till drawer should agree with it —less anything "Paid Out", i.e. bottle deposits refunded, etc., etc.

There is no reason whatever why any till should be other than exactly correct when checked, providing it has been rung properly, the right money taken and the right change given. Any variation is a reflection on those using it.

If for any reason some person requires money out of your till (petty cash, payment of bills, etc.) you should request a chit stating the amount taken. If one is not forthcoming you should put one in yourself, stating the amount, when taken, whom by and what for.

It is very upsetting when the "Guvnor" tells you that your till was £1.00 short the day before. Sometimes it is difficult to recall that the "Missus" borrowed it for cat's-meat in the morning. Guard your till and the money in it carefully.

NOTE: (1) Don't wait until you have run *completely* out of change before requesting more. *Before* you need it go and get it. This sort of thing slows down the service, annoys and sometimes loses customers.

NOTE: (2) Keep your till drawer tidy—with everything in the right divisions. This makes for speed and cuts out mistakes.

NOTE: (3) If your till becomes choked with money ask for it to be removed before any unauthorised villain decides to remove it for you.

Never on any account hold money in your hand for one order while you complete another one—even though you do eventually ring the two amounts later, either singly or together. Always, repeat *always*, ring each order at the time the cash is taken, with no dilly-dally, either!

The following piece of trickery is well-known and more dishonest barmen get caught over it than almost any other thing, so it is always being watched for. When a customer hands over the correct money (so that there is no need to go to the till for change) the barman holds on to it whilst he serves another order and then "forgets" to ring it up (but doesn't forget to slide it craftily in his own pocket!).

Fortunately for everyone they are always caught, some sooner than others, because every customer is a potential detective and many are personal friends of the "Guvnor" and "Missus" and are not going to stand by and watch them being robbed by a fiddling barman.

14
THE CROOKS—AND SOME
TRICKS TO CATCH YOU

In most public-houses some customers are "regulars" and after a short time you will begin to recognise them. This is perhaps not so important as recognising strangers who may be up to no good.

If two strangers are found in the bar at the same time, and have taken up separate positions, be very much on your guard, more especially if *one* of them engages you in close conversation—the other one may be up to a little "mullarky". Anything portable is fair game to public-house crooks, the Blind collecting box, the lighter fuel box, the Christmas stocking, the Spastics Beacon (even chairs and tables!), anything they can lay their thieving hands on. So give your eyes a treat. They may jump the bar and collect the "float" if the chances look fair for a quick getaway. This is a common and almost daily occurrence.

Often these gentlemen will ask you for something which is not readily available and when you, all unsuspecting, have disappeared to find it the coast is clear for them to go to work. So—*never leave your bar* without advising someone first.

ARTFUL DODGERS

[The following notes from the first edition of *Bar Service* are still valid, even though the examples are in old currency.]

There are several tricks designed to catch you for ten shillings and more. Unless you know them you are at a decided disadvantage against men who earn a comfortable living "on the fiddle". It is better to be prepared for them than to learn from costly experience—especially as suspicion may fall on you if your till is found to be short.

Trick No. 1. A man enters the bar when it is quiet and calls for a drink. He pays with a ten-shilling note; as he drinks up he puts his hand in his pocket and rattles some money. Calling the bartender over the conversation proceeds as follows:

Man: Excuse me! Did I give you a pound or ten shillings?
Barman: Ten shillings, sir!
Man: Funny thing—I thought it was a pound.

It now depends on Barman's insistence or uncertainty whether Man proceeds to insist it was a pound and hopes to receive another ten shillings change, or whether, because of Barman's absolute certainty, he proceeds with the other part of the trick.

Man: You're right, old man; It *was* ten shillings! I'm so sorry! All make mistakes, y'know! Sorry!
Barman: That's all right, sir, but I am certain!
Man: How stupid of me! I really didn't want to change that ten shillings! I've got a pocketful of change here. Would you mind letting me have it back and I'll change up some of this silver.

Barman, now quite relieved that he is not involved in a dispute, agrees and hands over the ten-shilling note. Man fiddles about with coppers and silver, meanwhile keeping up a cheerful conversation. Finally he places ten shillings in change on top of the ten-shilling note and says:

"That'll do! Give me a pound note for this, please!"

This trick has been worked hundreds of times with the barman not realising anything is wrong until the till is cashed up at night—and sometimes not even then!

The only way to foil this trick is to hold on to your 10/- note (you've already given him change for it, remember) until dodgy man has made up his mind exactly how much money he wants changed.

Trick No. 2. A man enters the bar and buys a drink, handing over a five-pound note for which he receives the correct change. He drinks up and vanishes. Next, another man, usually well-dressed and well-spoken, buys a drink and pays with a *one*-pound note. In a minute or so he calls the barman over and disputes his change.

He is quite certain he handed over a five-pound note and only received change of a pound. During the dispute, and to clinch the matter in his favour, he will say:

"I am absolutely certain about it. I went to the Bank and cashed a cheque for fifty." Here he will fiddle in his wallet and bring out nine brand-new fivers.

"I'm afraid I can't tell you whether I took your one from the top or the bottom, but the number must be either ——— or ———. Just have a look and see if it's there." Of course it is! The first man (his confederate) handed it over when he bought his drink. So once again you have been caught—unless you are right on your toes!

Trick No. 3. This is worked at a busy time. A man standing at the bar waiting for service has a five-pound note spread out on the counter in front of him. He passes some remark about it to the man standing next to him—a complete stranger probably.

"Don't go far these days, do they? What with cigarettes at five and five and Scotch at two-pounds eight and six a bottle, five pounds goes nowhere!

The stranger agrees and the man has made certain he has drawn ample attention to the five-pound note.

As he is being served, however, he switches the fiver for a folded one-pound note. A little later he will insist to the barman that he handed him a five-pound note but only received change of £1. He calls on the stranger as a witness and he, of course, affirms that he *saw* a five-pound note handed over—which, of course, he didn't! Heigh, ho! another four quid up in the air!

Trick No. 4. Do not accept a cheque from anyone unless you have permission from the person-in-charge, no matter what sort of tale the customer puts up. Your answer is that you have no authority to cash cheques for anyone.

The villain is usually clever enough to watch points very carefully before trying this on and he is probably planning to do the same thing in several pubs about the same time.

He waits and watches the movements of the Licensee. A little casual conversation with a foolish barmaid soon lets him know when the Guvnor goes out. Then comes his big moment: he walks in the bar telling the tale.

He has just met the "Guvnor" in the High Street and asked him to cash his pay cheque. He does it every week, y'know. The "Guvnor" told him to bring it in to you and said you would cash it because he hadn't enough on him. It's quite O.K. Nuisance this wages by cheque business, etc., etc.

The cheque is probably made out for forty-four pounds eighty-nine pence—and here comes the disarming part. He doesn't want it all—just let him have thirty pounds because he's got to pay the rates this morning and he'll collect the rest tonight when he sees the Guvnor!

Don't fall for it! Just tell him he'll have to wait till the Licensee comes back. At this there will be a display of temper (on his part because his trick has failed) and he will say he is in a terrific hurry. He is! He certainly won't wait until the Licensee returns and you will never see him again, so great is his "hurry".

Once in ten years the story may be true and the Licensee and the customer may be quite annoyed. Dismiss it as stupidity on the part of the Licensee and certainly don't let it worry you. You would have more cause to worry if you *had* parted with the money to a fraud—and ninety-nine times out of a hundred that is what it will be. If by chance the Licensee is particularly upset about your refusal to change what he knows to be a perfectly good cheque ask him quietly if you have his authority to cash others on his behalf.

Whilst on the subject of cheque changing you will come across the man trying hard to cash one whose friend standing by says—"If that's a dud—I'll cover it!"

The answer is—Well, why don't you cover it now while it's still good?

If you are authorised to change a cheque for regulars, whom you can recognise as such without referring back to the "office", always examine it.

1. See the date is right. It is in order if it is dated *before* the day on which it is being cashed (unless it's *months* old)—but if it bears a date in advance of this it is said to be "post-dated" and will not be accepted by the bank until the date shown.

2. Always examine every cheque for any alteration. *Every* alteration *must* be initialled by the person writing the cheque.
3. See that the figures and words agree. The words may read "Payable to —— Thirty Pounds only and the figures may say £3.00—obviously an error, making the cheque void until it has been sorted out and corrected. Don't pay out on it.
4. If the cheque involves a third party, if it is "drawn" by Mr. Smith in favour of Mr. Jones and your "Guvnor's" name is Mr. Robinson—then Mr. Jones must endorse it (sign it on the back). Cheques must always be endorsed in *exactly* the same way as they are drawn. For instance, Mr. Smith may have made a mistake in spelling—perhaps he has written F. JOMES instead of F. JONES— then Mr. Jones must endorse it "F. Jo*m*es" whether it is his name or not.
5. Make sure the cheque has been signed.

Trick No. 5. Don't get caught over "the Guvnor says" trick over drinks.

"The Guvnor said we could have a round of drinks till he comes back."

Don't stand for it. Ask them if the "Guvnor" gave them the money, then apologise and say you are sorry, you do not take instructions from anyone but the Licensee himself— and let them pant.

Trick No. 6. This concerns stock. A lady, with a gentleman seated in a car outside, enquires if she may have a bottle of White Horse Whisky. She is served and pays for it. In seconds she is back—she made a mistake—her husband wants Black and White—could it be changed—sorry to trouble you!

Later on it is discovered that the White Horse you put back on the shelf contains—you've guessed it—water! If this fraud does not come to light before you sell the "phoney" bottle then you may be in double trouble with the customer who buys it in good faith—as he may trot it down to H.M. Customs and Excise for analysis.

Trick No. 7. This is a variation on No. 6 but in a bigger way —and usually about Christmas-time. A customer confidentially whispers to the Licensee that he has had a bit of luck in a raffle and won a case of Whisky. (A *case* of whisky is an unusual prize anyway—maybe a bottle!)

However, he doesn't drink whisky—nor do his friends (here he produces a genuine bottle of proprietary Scotch and asks the Licensee to try a drop). He has a beer, himself. Now if the Licensee would like the other eleven bottles he can have them for a pound a bottle, as he'd rather have the money. If a deal is made then the Licensee has bought himself eleven nice bottles of water—No! Not again? Yes—it still works! Often the Licensee tries to pull a smart one over this and offers Chummy *ten pounds* which is reluctantly accepted.

However, he is well and truly caught because there is nothing he can do about it. It is illegal to purchase spirits from anyone other than a Licensed dealer—so he may be reluctant to inform the Police.

There are more variations of these two tricks. The bottle of Gin "left over from a party", the three boxes of cigars "found surplus at stocktaking", the Brandy "left behind on the furniture van"—all means of getting rid of rubbish, or if not rubbish, stolen property.

Trick No. 8. The "hurry-up" trick. A man rushes in the bar and orders a double Scotch. Then he will say—"Give me five ones for this quickly—I've got a taxi waiting outside and I'm in a shocking hurry." Don't let him fluster you—the "fiver'" is a dud!

Postal Orders

You should be as careful with Postal Orders as with cheques. Every Postal Order is marked *"Not Negotiable"* which means that if it is "negotiated" (or dealt) through a third party or more, the last *known* person to handle it would become liable if it happened to be stolen.

Trace one through. Jones buys it and without filling in the name sends it to Smith. Smith changes it with Brown. Brown passes it on to Green and Mr. Green, a regular of yours, changes it with Mr. Robinson—your Guvnor. Months later it is discovered that Jones didn't buy it at all—he stole it. So the Postmaster-General writes to Mr. Robinson's bank and besides telling him it was stolen tells the Bank they must make good the amount. The Bank, who don't stand much of that, advise Mr. Robinson that *he* will have to lose it. Mr. Robinson passes the buck on to Mr. Green—but

Mr. Green doesn't know Brown, or where he lives or where he works. So Mr. Green is the loser. Only the purchaser and the person to whom it is made payable are entitled to deal with a Postal Order.

Where a Postal Order (or a cheque) is crossed, i.e. when it has two lines across it and the words "& Co." written between them, then payment will only be made through a bank—although in the case of Postal Orders they could be paid into a Post Office Savings account.

The risk with Postal Orders lies in changing them for strangers. If you know the person and where he lives, then you will be able to take steps for recovery of the money if anything goes wrong. If *you* are the last traceable person to handle a stolen Postal Order it could at best put you to unnecessary trouble and at worst involve the Police.

In general practice you will find that many people accept the risk of changing them—but it is as well you should know the risk is there.

General Notes

Watch for the gentleman, perhaps not too well dressed, who walks about with an umbrella or walking-stick. Sometimes these have a handy spike on the end and can be used for spearing cigarettes off a shelf behind the bar while you are not looking.

Sometimes paper money is kept in a glass beside the till on an adjacent shelf. Make sure it is not in a mug with a handle—because the same umbrella or walking stick can be used to hook it up.

Never put a folded note in your till—always open it out before giving the change—however busy you are. It is well known for notes to be cut in half (a race-course trick) and this makes them go twice as far!

Never fool about with money—keep well away from it. Some very uncomfortable times have been had by bar-staff larking about with heaps of money. Sometimes there is a shortage which is nothing whatever to do with the person playing the fool, but suspicion will often rest there.

A stupid statement like—"I'll just take this little lot for my holiday"—may appear very funny at the time but can finish up as an "embarrassing moment".

Never hand back any money—unless you are in no doubt whatever that you have made a mistake. Your answer should always be: "I am very sorry but there is nothing I can do about it until the Till is checked". Then call the person-in-charge to deal with the matter.

Do not carry your own money in the bar. In some houses (and particularly all company houses) it is not allowed under any circumstances. You should not require any money while you are on duty. The rattling of loose change in a jacket pocket, even if it is your own, often gives rise to unwarranted suspicion.

You *must* make it an absolute rule never to lend any money to customers. Once you start it is very difficult to refuse—so don't start. Your answer to all requests for a loan should be, "I am not allowed to lend money to customers—or to borrow from them!"

If they keep "having a go" at you to buy them a drink you might point out that the board outside says—"Licensed to *sell* beers, wines and spirits". It says nothing about *buying* any for customers or giving any away, either.

15
THE LAW AND YOUR JOB

CHILDREN

Although it is not common knowledge (and Licensees do not publish the fact) a child *over fourteen* years of age commits no offence by being *in* a bar, but, of course, is not allowed to drink anything alcoholic. Licensees do all they can to discourage anyone under the age of eighteen being in any bar for fear that the child may be served, or given alcohol by an older person. This is a very serious offence and recently magistrates and Chief Constables have been giving the Trade some serious lectures on under-eighteen drinking. *Never take a risk* if you are in doubt. You are within your rights if you refuse. You are in trouble if you unwittingly serve anyone under age.

A child of *sixteen* years or over may be served with beer, cider or perry with a meal in a restaurant, or that part of the house usually set aside for the service of meals. If the house has a bar which is usually laid up for lunches during which time drinking without a meal is not allowed (even though at night it is used exclusively for drinking) then he or she may be served *at lunch* with beer, cider or perry from the age of *sixteen*.

Apart from this NO person under the age of eighteen years may be served with alcoholic liquor in any bar—*or outside any bar* nor in any off-licence for taking away.

Any person who *sends* anyone under eighteen for alcoholic liquor is committing an offence.

If any person attempts to obtain alcoholic liquor either for drinking or for taking away by any person under eighteen he is committing an offence. It is the Licensee's job to ensure that no alcoholic liquor *reaches* anyone under eighteen and *you* as his servant will be held equally liable if it occurs through you.

A drink on a table in front of a person is considered to be their drink (even if it isn't!) and in the event of a visit from the police it may be sampled and a prosecution result. It is difficult and expensive to refute corroborated police evidence. This is why Licensees always frown on anyone under eighteen years being in the bar at all.

It is no offence for a child to walk through a bar to, say, the toilet, if there is no other way to reach it (accompanied by an adult), nor for the children of the Licensee to be in the bar, or any other children who are resident on the premises. Young persons under eighteen may not in any circumstances be employed about the business of a bar while it is open.

You must be very careful about young children brought into the bar in secret and hidden behind partitions and curtains while their parents stand drinking. *This must be stopped.* If there is the slightest argument about it call the person in charge *at once*.

There have been cases where the police have threatened to apply for a "Care and Protection" order against obstinate parents stupid enough to argue. Sometimes you are entitled to remind a customer who objects to observing legal restrictions that Licensees, you, and policemen do not *make* the law—but must ensure that it is observed.

Don't forget that it is illegal to serve anyone under sixteen with cigarettes or tobacco even if they tell you (and you have reason to believe they are telling the truth) that they are fetching them for an adult.

DRUNKEN PERSONS

It frequently happens that a person who has taken on too much enters the bar and asks for service. *Do not serve him* until you have (quietly) seen the person in charge and taken instructions. It requires long experience in public-house psychology to assess the situation correctly and decide upon the appropriate action. Sometimes a blunt refusal is the signal for a load of trouble with smashed glasses and damage to property if not injury to the person. In these cases it is as well to stand prepared.

The police may, if called, prosecute the offender, but one must always take into account the question of "Drunk on Licensed Premises" which is a very serious offence for the Licensee. If there is sufficient proof that the person was drunk when he entered *and was not served* it may be safe to call in the police. Otherwise, as in far too many cases, the Licensee may just "grin and bear it" to avoid any trouble between himself and the "Law".

It is an offence to *supply* any alcoholic liquor to a drunken person even though he or she is not actually served but obtains the drink through someone else who is sober.

It is often very difficult to detect when a person has had "enough". A man may stand normally at the bar and in clear-cut tones ask for a pint of bitter. After being served he may stagger; he is quite hopelessly "under the influence" but this has not become apparent until then.

In such a case, there is only one thing to do: await an opportunity to get the beer back over the bar, refund the *full* amount of what he paid for it (whether he has consumed any or not)—and request him to leave. If you are wise you will have removed everything portable from within grasp—syphons, ashtrays, bottles, glasses, etc., etc., *before* any action is taken in case he starts throwing.

Refusal to Leave

It is an offence for any drunken or violent or objectionable person to refuse to leave the premises after having been requested to do so—and he may be ejected provided no more force than necessary is used.

Provocation and Violence

However much provocation is suffered by you it is unwise to offer any violence to a customer. Magistrates for years past have taken a far too lenient view of assaults by customers on people serving behind the bar, treating them as mere occupational hazards, yet the perverse working of so-called "justice" makes a rare "to-do" if the position is reversed.

Refusal to Serve

Few trades are subject to more pettifogging legal restrictions and regulations than the Licensed Trade. However, it *does* enjoy, in common with every other shopkeeper, the right of refusal to serve.

Just as you cannot *demand* anything in a shop window (whether priced or not), the person in charge of a public-house may refuse to serve anyone he chooses. However, persistent or unreasonable refusal to serve refreshment (not necessarily alcoholic liquor) may lead to trouble with the Licensing Bench. If you feel you must "refuse" NEVER give a *reason*—it may involve you in an action for slander, or worse.

REMOVE EVERYTHING PORTABLE

Never be belligerent when refusing to serve a customer. It is enough to tell the person concerned quietly that the "Guvnor" would prefer not to serve him or her. They should then be requested (politely and quietly) to leave the premises. If they refuse to go they may then be ejected on the grounds of trespass.

BETTING AND GAMING

In spite of what anyone may tell you, Betting and the writing of Betting Slips still constitute serious offences on licensed premises.

Many people are under the mistaken impression that the Licensing Act of 1961 legalised all forms of betting and gaming in pubs. Although Betting Shops are now permitted and some concessions have been made in respect of gambling the position with regard to public-houses is not much changed.

It *is* permitted to play games of skill which includes shove-halfpenny, darts, billiards and skittles for money—without restriction as to stakes. Dominoes and the card game of Cribbage are allowed, but only for small stakes, and what constitutes the legal definition of *small* apparently still awaits a test case. So Solo, Brag, Nap, Pontoon, Banker, Poker, etc., etc., are still illegal.

PERMITTED HOURS

Generally, "Permitted Hours" total $62\frac{1}{2}$ per week—exclusive of drinking-up time.

The usual hours for public-houses in London are $9\frac{1}{2}$ per week-day and $5\frac{1}{2}$ on Sunday. (There may be some variation of hours in certain parts dependent upon the district. Market towns sometimes have special hours on market days which are known as "General Order of Exemption".)

The hours usually are:

11 a.m.	to 3 p.m.	= 4 hours	Week-days
5.30 p.m.	to 11 p.m.	= $5\frac{1}{2}$ hours	
12 noon	to 2 p.m.	= 2 hours	Sundays
7 p.m.	to 10.30 p.m.	= $3\frac{1}{2}$ hours	

plus ten minutes' drinking-up time after each session. Within prescribed limits hours are at the discretion of the local Licensing Bench.

A Licensee is not *forced by law* to open at all, in spite of

147

the loud-mouthed individual banging on the bar door before you open. Three considerations have a bearing on opening time.

(a) A Licence may become forfeit if interested parties (Salvation Army, Nonconformist Church, militant temperance bodies and local Licensees) are able to prove that there is no public need for any particular house on the grounds that it does not open the *full* permitted hours.

(b) In the case of "Tied houses" the Brewers insist that the Tenant shall remain open the full permitted hours. Failure to do so would be a breach of the tenancy agreement.

(c) There may be a stipulation in the Licence insisting on full hours being observed.

Drinking-up Time

This is the ten minutes directly following "Permitted Hours" at each session. No drinks may be served during this period.

This is a new provision in the Licensing Act 1961, which (at last) removes the absurdity of a person being allowed to order and be served with a drink one second before time and being liable to prosecution for drinking it.

In the case of a person who is having a meal and who has been served with a drink before time the Law allows 30 minutes' drinking-up time. Such a person may not be served with drink after normal "permitted hours", however.

After Time

Serving after time is a serious offence. Generally the police will be found reasonable with regard to this. They rarely take action over a couple of minutes—realising that when they hear a till ring it may be the completion of an order commenced well before the second bell, or someone buying a bag of crisps.

However, any blatant or persistent breach of this regulation will certainly merit a call from the local inspector if not a summons.

You will find it best *never* to serve after time—not even half-a-minute.

Sometimes it is very difficult to refuse. One of your regulars may have galloped a mile to get a drink, not having had one all night, and arrives panting in the bar just as the

second bell stops ringing. Your best course is to see the person-in-charge for after all you have no right to jeopardise the licence. Remember that if you serve *one* person after the last bell others will demand service as well.

An "After Time" Trick

Don't be caught by one crafty trick of the customer with a party who, having been refused a round of drinks after time, shows some reluctance to move on. Coming to the bar again he will appear most concerned and say apologetically:

"I'm *so* sorry—I haven't bought Alice (the barmaid) a drink all the evening. Will you please ask her what she'll have—and have one yourself!

Don't fall for this. Politely refuse. If you don't you are in a very awkward position when, before paying, he says— "And I'd better have three large Gins and two pints of Bitter —and what'll you have, Charlie?"

The drink for you and Alice is just his entrance fee for another round of drinks after time.

WHO MAY BE SERVED

Any person *resident* on licensed premises may be served

149

with drink *at any time* as may any private friends of the resident who are being genuinely entertained by him *at his own expense*. The friends are not allowed to purchase drinks themselves out of hours.

A customer may not be turned into a friend for the purpose of evading the law. The police will soon expose any such subterfuge by separating the persons involved, taking personal details from each and comparing notes.

CREDIT

Credit may not be given for drinks consumed on the premises. If any item is taken as a pledge for spirits supplied, i.e. if a man leaves his watch with you as security for a couple of double whiskies, the watch may be reclaimed and the Licensee fined.

In the case of beer or cider supplied on credit there is no method of recovering the money owing.

Credit may be given in the case of goods delivered from an off-licence and civil action may be taken to recover any money owing on an account.

OFF-LICENCES

No drinking is allowed in any "Off-licence" whether it is attached to a public-house or not.

The words "Off-licence" mean premises licensed for the sale of alcoholic liquor for consumption *off the premises*.

A Licensee has a duty to see that no person or persons buys drink in an off-licence and takes it outside to drink. For instance, if he buys a crate of beer and sits down outside the premises to drink it the Licensee becomes liable.

The police frown on *all* drinking outside licensed premises and may insist that it is stopped.

MEASURES

Legal points regarding measures and glasses have been fully explained in the sections on Beers, Wines and Spirits, etc., but it is as well to repeat that draught beer served in quantities of half-a-pint or more *must* be in Government-stamped measures. Wines and spirits dispensed by the half-quartern (half-gill) or quartern (gill) must also be served from stamped measures—and then poured into the glass. The Weights and Measures Department will order you to remove *all* measures which are damaged or dented.

SMOKING

It is an offence under the Food and Drugs Act 1955 to smoke or take snuff behind the bar or in any place where food is kept or exposed for sale—liquor being classed as "food" under the Act.

This apparently applies at all times, whether the house is open or closed.

A recent prosecution resulted in a fine for a Licensee who was found smoking behind the bar when his "pub" was shut.

The rights and wrongs of this piece of legislation may appear a little confused when it is considered that no parallel order forbids coughing or sneezing into a customer's beer nor a customer standing inches away blowing smoke into the Guvnor's or a customer's beer or sandwiches.

In fact the trouble stems from contamination by the mouth. The smoker, in handling a cigarette, will come into contact with the end which has been in his mouth. Germs are therefore conveyed to the hands and from thence to glasses, cloths, washing-up water, etc.—perhaps to food.

COMMON LAW ON DISMISSAL

The Common law (not the Catering Wages Act) provides that no wages are due to any person dismissed through:

1. Theft, embezzlement, being drunk, rudeness, insolence or misconduct.
2. Disobeying proper legal orders.
3. Being incompetent.

Example: A Barman having worked five days in a week, and who is properly dismissed for any of the above, *cannot claim any pay* for the five days already worked.

Leaving without proper notice

If you wish to give notice to your employer the proper time to do so is at the time of payment of wages. Failure to give proper notice renders an employee liable to an action for damages—the damage usually being assessed at the amount of the wages involved plus costs, of course!

BAR-ROOM LAWYERS

Legal experts in the bar—usually fully paid-up dustmen—will talk a lot of rubbish to you concerning Licensing law. Smile indulgently, but don't become involved. It is ten to one against them knowing law that even lawyers have to wrestle with. Some people seem to think that because it is commonly known as a "public" house they are entitled to do as they like within its walls. It would be nearer the point to say it is a "Private" house—open to the public at the discretion of the Licensee, the Bench—and the brewers, if it happens to be "tied".

16
STOCKS AND STOCKTAKING

In Managed houses, stocktaking is a regular event and everyone stands or falls by the result—from the manager himself down to the learner-barman.

Every effort must be made on your part to see that nothing is done to cause stock shortage.

Your cigarettes *must* be paid for. It is far safer and wiser to ask someone else to secure them for you—someone who can verify that they are paid for. Even better than this is to smoke a brand not sold in the house.

Licensees are not "mugs"—they are shrewd men. They have to be.

It is not the job of *this* book to explain to you the many ways of trapping light-fingered bar staff, but suffice it to say that they exist, and that the offenders look particularly stupid when they are caught and often lose a great deal more than they gain, including their character.

Bar staff rate 85 per cent on the count of strict honesty, the other 15 per cent being divided between the attributes of cleanliness, quick service, pleasant temperament, punctuality and accuracy.

If by any chance you should notice anyone doing anything that would affect the stocks such as:

Drinking without paying;

Passing drinks over the bar without accepting payment;

Ringing up less than the total order and holding on to the balance;

Taking cigarettes without paying;

you are quite justified in taking the matter up with them on the grounds that you are all equally responsible for the stocks and will *all* get the blame for shortages. Any person indulging in these practices is not worth working with anyway.

17
SOME GOLDEN RULES OF
CONDUCT AND PROCEDURE

IN THE BAR

Some of the points covered here have been dealt with earlier but they are repeated again for additional emphasis.

Any idea that you are entering the Trade for the sole purpose of enjoying yourself and being paid for it might as well be dispelled right now. It is hard work—go! go! go! but nevertheless, it is a job which carries a great deal of personal satisfaction, especially if you are qualified in all branches and are able to answer questions, and give advice, with authority. It is amazing how much ignorance prevails in the Licensed Trade, from "Guvnors" and Cellarmen to Barmaids and Customers.

By making the Trade your interest and hobby instead of just your living, you will soon be surprised how many people will bow before your superior knowledge. But, please! don't flaunt it and become known as an unbearable Bighead!

ENQUIRE ABOUT THE DOG

Before you go behind the bar make certain you enquire about the dog (if any) and where it is kept—and if it is chained. A surprise meeting will probably shake you—and the dog!

Enquire about your till. Some houses leave the till roll on until it runs out, others remove the roll each day. Some reset the total each day, some once a week, some after each session. Make sure it reads zero when you start. If it doesn't, ask why? Zero on most tills is a row of six noughts.

Watch for the till roll changing colour—turning red. As soon as you see this you will know it is running out and requires to be changed, so mention it as soon as it is noticed.

Do not run out of change. As soon as you are running out ask for more. *Do not* take change from one till to another —buy it!

It may be your responsibility on occasions to take money from the Office. Here is something most important: if, for instance, you have to take five one-pound notes in exchange for a "fiver" and the only pound notes are those bundled up with a paper band indicating that the bundle contains £100 —always remove or break the band. Hours have been wasted at closing time because a bundle thought to contain £100 actually only contained £95, through someone neglecting to tear off the band.

Do not eat anything behind the bar (or chew gum) during opening time. It looks bad from the customers' point of view. You meal times are fixed by the C.W.A. (Catering Wages Act) so there is no excuse. Leave the cocktail cherries alone—they cost money. Although they may look tempting don't make your lunch off them.

If someone from the Kitchen is kind enough to give you a cup of tea in the bar, have the decency to:
1. Wash up the cup and saucer.
2. Return it to the Kitchen.
3. Thank the Cook!

Do not be late down in the bar—especially at "opening

time". "Punctuality is the Politeness of Princes" says an old motto, and it shows a sense of duty if you are always five minutes before time. This will allow for any emergency to be dealt with—beer not drawn up, floats not out, or perhaps a blown fuse.

If you *do* happen to be late, have the decency to apologise —and give the *correct* reason. Don't make up a pack of lies. The truth will usually be accepted as such—even if it does not go down too well at the time.

It is YOUR responsibility to get up in the morning— nobody's else's. So do not try making someone else responsible for getting you out of bed. Buy an alarm clock.

Don't treat your job as just an unpleasant way of earning a living. Try and develop a sense of responsibility. Look upon everything as yours—*your* bar, *your* glasses, *your* customers, and take a pride in your equipment.
A few minutes spent on that filthy copper funnel with the metal polish will do you credit—and not go unnoticed by the Licensee, or the customers.

Keep an eye on little things—such as turning out the dartboard light when play is finished.

Watch for "opening" and "closing" time. If it looks as though you are running "after time" a hint to the person in charge may not come amiss. It is just possible that it may be resented at first—*but*—in time it will be realised that you, at least, are not asleep.

Master the electric light switches. You may find yourself alone in the bar and need to switch on—so find out where they are and which is which, as and when the opportunity occurs. If the lights should go out suddenly, stand by your till!

Never leave your bar unattended. If you *must* leave it for any reason, always make sure some other member of the staff is told of your intended absence. Your place is IN the bar—not outside it. Even if the Beatles *are* playing on the pavement opposite—let them play and get on with your work.

Do not wander about from one bar to another (unless on business) even if there happens to be nobody in your bar at the time. It is more than likely that the very moment you leave someone will come in, find no service, and be kept waiting.

Never leave your flap open (unless it is kept that way permanently). This applies whether you are behind the bar or not. It is not safe to do so. If it locks—keep it locked. There are many sneak-thieves about just waiting the opportunity to slip behind your bar for no good purpose.

Keep a sharp eye on tramps, dirty-looking people, hawkers, or anyone with an obvious disease. Inform the person in charge if anyone in your bar appears to be annoying your customers.

KEEP AN EYE ON TRAMPS

Never serve any person who is drunk. It is an offence in law and the penalties are severe—both for you and the Licensee.

Make sure you know *in advance* what to do in case of fire.

157

Do not serve any person who walks in and asks for fourteen double Gins! There is usually something wrong here—so see the person in charge.

It is illegal to smoke behind the bar—*even when the house is closed!* It is illegal to take snuff behind the bar.

Keep the customers' ashtrays clear—and clean.
Make absolutely certain no cigarettes are burning before you throw them in the rubbish bin. Many serious fires have been caused this way.

Do not grab the "Guvnor's" paper as soon as it arrives, open it out to find the winner of the 4.15 and then, when you've read all you want to, let him have it—well crumpled up. This is a piece of impertinence. If you want to read a paper—go out and buy one, or wait till someone has finished with it.

Even if there are no customers in the bar, do not spread a newspaper all over the counter to read.

If drip mats are used make sure they are always clean. They are supplied in great quantities by many firms free of charge as an advertisement, and there is no excuse for dirty ones to be in use.

Make sure a soda syphon is readily available and that it is not empty.

Always give each new syphon a short squirt in the sink before using it. This is to release any excess gas.

Make sure you have the following ready: Ice Cubes, Lemon Slices or Lemon and Knife, Cocktail Cherries, Cocktail Sticks.

Make sure your water jug contains *fresh* water. Yesterday's water may *look* all right—but it could spoil a customer's whisky.

Watch the dartboards (if any). They require regular soaking. Elm boards require a twenty-four-hour soak every two or three days otherwise they will dry out, shrink, deteriorate very rapidly and the wires will buckle. They are then useless.

Have a cloth and chalk ready for your dart players.

Although it may be nothing to do with you, see that soap, towels and toilet rolls are available in the bar toilets if these items are usually supplied, or else report to the person in charge.

Do not use the telephone without permission—and always pay for the call.
If a call comes through for you during permitted hours, keep it short and sweet.

A notice on a certain factory wall reads:
"If you have nothing to do—don't do it here!"
This is particularly apt for the public-house. There is *always* something to do—dusting, polishing glasses, etc., etc.

Do not stand chatting and be unaware of the presence of customers awaiting service. Customers come first; serve them *immediately*—the chat can wait.
Always recognise your customers—even though you may be busy serving—it is annoying for them to be ignored whilst someone tells you what happened on the Somme in 1916.
Avoid those customers you know will want to engage you in conversation for the rest of the session; be pleasant but make yourself busy.

Tidiness is an essential part of good bar-keeping. It is bad if, when you want to cut up a lemon, a ten-minute search reveals that the special knife usually kept for this purpose in the lounge cabinet is in the public-bar, having been recently used for digging nails out of the dartboard! Similarly, if YOU use anything put it back in its right place, where it can be found by the next person.

It is one of the time-honoured features of the English public-house for the "regulars" to have a bit of fun with the bar staff—especially with pretty barmaids. You will be expected to take this in good part—and even join in. The purchase of a Brown Ale, however, does not entitle anyone to take liberties and you should see that the conversation never degenerates below the level of propriety. If you find that someone is constantly being objectionable, or over-suggestive, it is correct to lay a complaint with the person in charge. Never take matters into your own hands, or lose your temper—just make a report.

Pick up any loose crown corks on the floor—don't kick 'em about. A crown cork can cause endless trouble if it happens to get jammed underneath a door.

Remember what was said about cigarettes.

Do not decorate your display cabinet with raffle tickets, pools coupons or correspondence.

A great deal of trouble has been caused through goods being taken over the bar for safe keeping. Often a customer will ask you to mind her shopping bag while she goes along to the Post Office. This may seem all right on the face of it, and looks very churlish if you refuse, but . . . her bag may contain some eggs and although *you* are careful enough, your great ham-fisted cellarman might break them. Now what happens? The customer looks at you. You look at the cellarman. He, of course, says "Well they shouldn't be there!"—and right he is.

It is far better to say you will take the goods over—but you accept no responsibility whatsoever.

Do not leave swabs, glass cloths, or any cleaning material on the counter. It gives a bad impression to anyone entering the bar.

Swabs used for wiping tables, the pewter, ashtrays, etc. must never on any account be rinsed in water being used for washing-up.

Keep your own swabs and cloths—and hide them. Cleaners and others have a handy habit of picking up any cloths they can find—for *any* old purpose. Make sure they don't get hold of yours!

Watch your glasses as you wash them—they are expensive and it would stand to your credit if you could honestly say "*I* have not broken a glass in . . . months."

A common cause of breakage (or chipping) is catching the glass on the tap over the sink. If you can secure a short piece of rubber tube to fit over the tap it may save many breakages.

Glasses should be *dried* on one cloth and *polished* on another.

If you should serve a glass with a smear of lipstick on it—be diplomatic! Take it back and say it was cracked—or something!

GOODS FOR SAFE KEEPING

Do not turn on the tap at your well, walk away and forget it. Bar sinks rarely have overflow vents—so a flood is practically certain.

Although customers and friends may call the "Guvnor" Harry and the "Missus" Ethel *you* always refer to them as Mr. or Mrs. Brown—or Guvnor and Missus—and *nothing else*. Even if they tell you to address them familiarly in private, don't do so in front of customers.

If anyone wants to know if the Guvnor (or Missus) is in—do not know! Ask the person for their card or name and say you will go and find out—even if you do know. It may be an unwelcome visitor.

On the telephone the same instructions apply. If you answer the ring don't say HELLO!—give your number loud and clear. When you receive your reply, always enquire who is speaking before passing on any message.

"Mind your own business." This cannot be too strongly stressed. What the "Guvnor" or "Missus" do is *their* affair —not yours. Never mention anything that goes on in the house (or anywhere else) or what was said *to anyone, by anyone, anywhere*. Just know nothing!

A clever solicitor once said—"It is my considered opinion that ninety per cent of this world's troubles are caused through people opening their big mouths." Keep out of trouble.

Any room marked PRIVATE means just that! Knock on *all* doors before entering.

Be careful how you address your customers especially if they happen to be accompanied by a stranger. You may know one as George and another as Mary, but unless you know the surname and address them as "Mr." or "Mrs." Jones just say "Sir" or "Madam". It is possible they may not wish the stranger to know they are a regular user of your house.

Never commit the unpardonable sin of calling anyone "dear" if they are in company. Many domestic squabbles have been caused by this.

If a customer leaves the bar and returns later, either alone or in company, don't say anything like—"Hello! you back again?" Just say—"Good evening, sir" (or as the case may be) as though you hadn't seen him since the week before.

If a telephone caller asks if Mr. So-and-so is in the Bar, don't say yes, even if you know he is. Say "I'll go and enquire" and try to find out the name of the caller. Then ask Mr. So-and-so if he will take the call. He *may* not want to be bothered by the caller when he's enjoying his pint!

Make it a point of concern and pride that no customer ever walks out of your bar disgruntled or dissatisfied, if it is humanly possible to avoid it. Treat them as your guests, with proper respect, and don't forget to wish them "good day" on *entering* and *leaving*.

Never run the house down—or the Guvnor, or Missus, the other staff, the customers, the brewers. It has a funny way of getting back.

Do not get involved in any argument in the bar. It is not your business to interfere in any discussion or row between customers—unless you are directly concerned. Keep well away.

Never repeat any rumours or bits of "news". There have been some mighty red necks through repeating that old Sam Browne was found dead in bed the day before—especially when he walks in the day after.

Avoid discussion of religion or politics in your bar. They are dynamite!

The drinking public from every sphere, you will soon discover, are the most obstinate, ill-informed and perverse section of the community it is possible to find. Even if they have asked a question they will often refuse to accept the answer—right though it may be. Afford them an indulgent smile and let them wallow in their ignorance rather than risk a row by trying to put them right.

Do not wager on anything silly in the bar. If a man wants to bet you five shillings that he can pour a pint of beer in one ear and out of the other, just point to the notice on the wall—NO BETTING OR GAMING ALLOWED.

The television set in your bar is there for the benefit of the customers—not you! The same applies to the radio and the piano. If anyone asks to have the T.V. or Radio turned on—ask the person in charge. There may be good reason for leaving them off.

Customers will very often ask you to have a drink. It is most important to discover right from the start:
1. If you are permitted to accept drinks.
2. If you are permitted to accept money in lieu.

In many houses staff are *not* permitted to drink at all during opening time.

In some they are given a "staff drink" when the session ends.

In others they are allowed to accept the cash equivalent— "and have it later with my supper".

Again, in some, the "Guvnors" "couldn't care less". But don't be silly—and don't be greedy. Even if the only drink you really like is Green Chartreuse it would be a gross liberty to ask for one—or even to mention it! No customer should be expected to pay much more than the price of half a bitter for your drink but, of course, much depends on the circumstances prevailing at the time.

DON'T BE GREEDY

Avoid carrying any money with you on duty—you should not need any. A pocket tingling with silver gives rise to suspicion which may be wholly unjustified.

Work for wages—not for wages and pickings.

A warning must be given to all persons starting in the Trade that there *are* customers who may invite you to act dishonestly. For instance, you may be puzzled by a casual remark:

"Have you started earning yet?"

By this he means—"Have you started handing out drinks and cigarettes either without payment, or for much less than the value of the order, and accepting a generous tip for the favour? The next step is to snatch a handful of notes from the till.

Should any such remark be made to you, pretend you have not "cottoned on", and then mention what was said to the Guvnor. It may be the means of uncovering a serious "leakage" and you may depend upon it that anyone who falls for this villainy will most certainly be caught.

Work for the Guvnor or the firm! This means—don't waste anything. It is truly surprising how small amounts of waste add up to considerable sums of money—far more than you would care to lose out of your wages.

Keep an eye on new Relief Staff. It is unfortunately true that a few of them come to see what can be made on the side. As *you* stand to get the blame equally with any villains who get behind your bar, protect your own interests by giving your eyes a treat!

If you come across a split copper or silver bag, tear it completely in half and throw it away, so that it cannot be used again. It causes a nice scramble (and wastes a lot of time) if someone drops five pounds of five pence pieces into a silver bag with a hole in the bottom. For some strange reason the most you will ever pick up off the floor is four pounds and 95p.

Never pour any drink back into the bottle unless you are 200 per cent *certain* it has not been contaminated—spirits especially. It is better, if you are *not sure,* to get a second opinion.

Always wash spirit measures after use. This is important. A little Rum, for instance, left in the bottom of a measure, will taint Vodka which has almost no flavour.

Turn spirit measures upside down after use.

Never throw bottles into bottle baskets. The risk of breakage is obvious and the danger of injury to the cellarman who has to empty the basket is serious. *Always* advise the cellarman of any broken bottle in the basket *before* he starts to empty it.

Drain all bottles into the waste before you replace them in their cases. If you think this is an unnecessary chore, listen to the basic English of the draymen if they receive a showerbath when throwing the cases up on the dray.

Watch all beer boxes and cases for loose ends of wire sticking out. Apart from nasty personal injuries these wires wreak havoc with trousers, etc. If you are unlucky enough to receive an injury yourself, make sure you have it properly dressed. If you receive a splinter make sure it is out—and *not* with a rusty safety-pin!

Do not leave Riders in beer cases. Riders are those bottles which are placed on top of the stipulated quantity. Bottles cost money—there is no point in giving it away. They will usually get broken and no credit will be given for them anyway.

If no waiter is employed ask your customers if they would like a tray when handing over a large order—and make sure it is clean!

When customers bring their own bottles in which to "take away" draught beer, it is a wise precaution (if the Guvnor approves) to set their bottle aside and use one of your own. In any case wash and *smell* his bottle before you fill it. If (perhaps unknown to the customer) someone has used it for paraffin, *you* may get into serious trouble.

If a special book is kept for entering up the "off sales" of bottles of wines and spirits *always* enter up the sale *immediately,* not the next day.

Never help yourself to a drink without permission. Even if you *buy* a drink it may leave a doubt in the mind of anyone who did not just happen to see the money rung up. Customers also tend to exaggerate—"Look at her—she's always drinking"—though that may be the first drink you have ever had behind the bar in your life.

AFTER CLOSING TIME

At closing time, i.e. when service finishes, commence to wipe down your pewter, rinse and wipe all dirty glasses.

Do not start screaming "Time" as soon as the second bell rings. The Licensing Act of 1963 specifically allows ten minutes' drinking-up time—so it is *not* time until then!

It is very important to have all glasses and bottles collected in as soon as possible after "drinking-up time".

Be exceedingly careful *never* to pick up a glass with anything in it. A small Gin is sometimes difficult to see, but the customer having paid for it, will be ready to start something if it disappears.

If your customers are slow in making a move, unemptied glasses in front of them may be taken by the Police as evidence of "consuming after permitted hours"—with a possible

charge of "aiding and abetting" against you, or the Licensee, or both.

After *all* the glasses are in, the next priority is to clear the bar of customers.

NOTE: It is not an offence for customers to be *in* the bar after hours, as long as there is no evidence of drinking, but many customers come *in* at closing time and think this entitles them to drink it out till opening time.

Customers who have overstayed their time and still have drink left would be well advised not to touch it, in the event of a visit from the Police, as in these cases there is absolutely no defence.

When the beer has been removed from your drip-cans wash them out and leave them standing *upside down,* wipe and dry your counter. Empty the water jug and leave it upside down (if this is possible). Replace all wine, spirit and squash bottles in their proper places. Replace all measures and other equipment. If no Potman is available (usually his job) grab a broom and make a start on the floor. As you go round, empty and wipe all ashtrays and wipe all the tables (not forgetting underneath the edges). Run a damp cloth under the front edge of the bar counter top.

Any money you find on the floor is *not* yours—so be careful to hand it over. Sometimes a coin which should be in the till is dropped on the floor and cannot be found immediately, but it still belongs to the till.

Small coins found on sweeping up the bar floor are often allowed to be kept by the "sweeper"—but it is never worth risking your character for 10p. Pass it over the bar. Obviously you will hand to the Office any large sums found (making a note of exactly where you found it) in the hope that it can be safely restored to its owner.

Any items of jewellery, handbags, cigarette lighters, umbrellas, even half-empty packets of cigarettes *must* be handed to the person in charge. Don't just put them on one side—hand them over immediately. Theft by finding is an offence—wherever the property is found.

If you are given the job of closing the toilets always LOOK in them first—male or female. It has been known for a man to try hiding on top of the cistern. Look behind all doors. Sunday afternoon is a favourite time for hiding in the public-house and then "breaking out" when all is quiet, with anything going.

Take a look under any large settees.

Be very careful if you put out any lights at closing time whilst customers are still on the premises. Accidents have occurred through people stumbling in a darkened bar. See your customers have sufficient illumination for their own safety.

Hang all glass cloths and swabs up to dry before leaving your bar.

It is not usual to close till drawers when closing the bar. The reason is that if there is a "break-in" the thieves will probably smash the till, valued at up at £500, to get at the £2.50 float.

When bolting doors *always* leave the knob of vertical bolts turned towards the hinges. There are unsavoury gentlemen who, by placing a knife through the doors, are able to open

LOOK BEHIND ALL DOORS

them by pulling down the bolts—if they are easy enough to reach.

Make sure all the *fanlights* are secure. These are often used for gaining an entry.

Make absolutely certain that the cellar flaps are bolted and secure. (Recently an epidemic of "breaking and entering" by means of the cellar flaps took place in North London. The thieves gained entry in each case by drilling holes in the wood of the flaps and withdrawing the bolts.)

When you have completed your chores don't cause friction by disappearing upstairs without giving a hand to any other member of the staff who needs it. By the same token you could expect them to help you. "Give and take" is a good motto.

It is often a sore point to see Relief Staff packing up and leaving before all the cleaning up is done. In many cases this is inevitable—they may have transport to catch, or be faced with paying for a taxi.

Such is the state of things in the Trade now, that the employment of Relief Staff is inevitable and the fact that they can be secured at all is something to be thankful for. So don't be too intolerant of them—even though you may feel imposed upon.

If you know the Stocktaker is coming try to straighten up your bottles so that they can be counted easily.

IN YOUR LIVING QUARTERS

As soon as possible after engagement find out the daily routine of the house, and make sure you know:

> Your times of duty.
> Any breaks.
> Toilet arrangements.
> Facilities for laundry.
> Your time off—and if it is a permanent arrangement.
> The address of a local doctor and waste no time in registering on his list.

Always keep your room clean and tidy. Open the window each morning and turn back the bed linen to air. A housemaid may be employed to attend to the rooms—if not, make your own bed.

Representatives from the firm may wish to inspect the domestic side of the premises, and will get a poor impression of you and your background if they find socks draped over the fireplace, or a pair of dirty slacks on the floor with the bedspread.

If you smoke in your room use an ashtray (not the fireplace) and don't forget to empty it!

Don't smoke in bed—it is a dangerous habit; don't smoke at all—that's a better habit!

Don't rig up any electrical apparatus in your room until you have obtained permission. Electric irons, fires, tape recorders, record players, etc., may fuse the lights and cause endless bother.

Turn off electricity when not required. Do not leave anything burning during the night. It is dangerous, it reflects very heavily in the bills—and it is a sheer waste.

Don't leave food about in your room to encourage mice. Tin boxes are cheap and effective.

Leave all toilet facilities clean—sinks, wash-hand basins, baths, W.C.'s, etc.—as clean as you would wish to find them.

Doors are made to open and close. This particularly applies to the Toilet. *Always* close the door after use.

Do not treat your public-house as though it were a private hotel and you a paying guest—and don't try giving instructions in the Kitchen about your meals. Most cooks are very fair about likes and dislikes but few of them are going to stand catering to a diet sheet.

Unless you are inviting sudden death, never raid the "fridge" or larder.

It is often the practice for the "Guvnor" and "Missus" to rest during the afternoon break. Many bar staff also rest. If you are blessed with overflowing energy, please don't use it during the afternoon break.

When the house is closed, any noise, probably unnoticed otherwise, ie greatly amplified, so do not disturb the peace by banging doors, drawers, or stamping up and down stairs and passes. Turn your transistor down—and be quiet.

If you require to attend Mass, mention the fact to the person in charge. It is usual to attend in your own time—not when it means leaving other staff to do your work.

If you have corns, prepare to shed them now! Your feet are very important and should be looked after regularly. A visit to a chiropodist often makes a wonderful difference if you have any foot troubles.

FEET ARE IMPORTANT — LOOK AFTER THEM.

If all this bewilders you, just remember that most of it is just plain commonsense—the sort of thing you, being well brought up, would do naturally. If some of the advice has offended you just because of this fact, think of the people you have met whom it *would* have benefited!

A very good thing for tired and aching feet is a soaking in warmish water to which equal quantities of ordinary salt and soda has been added—enough to fill a teacup.

Watch your health and don't neglect yourself. The demands of the Trade are such that good health is essential. See your doctor before any complaint becomes serious. If you haven't registered with a doctor—do! Likewise a good dentist. Bad teeth means *bad breath,* and that should disqualify any bar staff.

Five minutes "knees bend" and general exercises will work wonders for you, if you do them regularly every morning and a short walk in the fresh air with the dog will do you good, and no doubt be appreciated by the Licensee—and the dog.

Every public-house has a bathroom for the staff. Make sure you use it regularly. Once a week is *the absolute minimum* for the enjoyment of health, and to remain within the bounds of decency.

A hot bath, with a couple of Bath Cubes added to the water, is a wonderful reviver after a long and tiring day.

Make certain your overall (or jacket) is spotlessly clean and that you yourself look clean and well-brushed. Keep your shoes polished.

Have your hands clean and nails trimmed. Remember you are handling "Food". Drink *is* food—by Act of Parliament!

Do not hang about on your day off—either in the house, or in the bar. Get out of it!

Some "Guvnors" object to serving their own staff who are off duty, so it is better not to try rather than be refused.

When you finally leave a post do *not* come back in the house, especially for a drink. In many cases this is not permitted anyway, and usually "ex-staff" are not in the house to do the Guvnor any favours—only chat about him.

APPENDIX ONE

GLOSSARY OF SOME OF THE TERMS USED IN THE TRADE [1]

Beer Licence, Beer House: Entitles the holder to sell *beer* only. It includes Cider and Perry, but *not* Wines and Spirits.

"Bubbly": Term used for Champagne.

Can: Generally means a metal tankard made of silver, pewter, or EPNS. Often kept on show for the use of one particular customer and may be his personal property. (When cleaning do not use metal polish on the inside—it may taint the beer.)

Canister: Canister, or Keg beer is delivered in sealed containers and is dispensed at the bar under pressure instead of being drawn up by pumps· This method saves much labour and represents the difference between the village pump and the domestic tap. A bottle of gas (CO_2) is connected to the canister (through a reduction valve and tubing) and the pressure forces the beer up

---

[1] This does not purport to be a full glossary, but will help those whose knowledge is limited. Certain other terms are explained in the main text.

to the service point through the beer service pipe which is attached to the top of the canister as required. There are various fitments for attaching these pipes, all very simple, and changing over from one canister to another is only a matter of seconds. A child could do it. The system has many advantages and is becoming general practice. On the question of whether canister beer is strictly "draught" beer, or not, one authority has stated definitely that it is *not*—it is bottled beer served from a larger container than usual. Others claim it is still draught beer. (*Note*: Some cans require only one pipe, the container having already been "gassed" at the brewery.)

Carafe: A container after the manner of a decanter much used in France for the service of beverage wines at table.

Case, Crate: A case is a container for carrying half-pint or pint bottles, crates being used for quarts.

Co-Licensee: A person holding a licence jointly with another.

Collar: A slang term indicating excessive froth on a glass of draught beer.

Collector: "The gentleman from the Brewery" who calls once a month to collect a cheque for the goods delivered. In many cases his duties do not now include "collecting" as accounts are settled on Statements rendered. However, he will make periodic visits to the house and is sometimes almost the only contact a Licensee has with his Brewery from one year's end to the next.

Condition: Is the "state" of any beer. "Too much condition" means it is "high" or gassy. "No condition" means it is "quiet" or flat.

Corkage: A charge made for "service" when selling a full bottle of wine to be drunk on the premises.

Corky, Corked: A term used for wine which has deteriorated through a faulty cork.

Cullet: Broken glass—sometimes collected for re-melting.

Decant: To pour from one vessel to another.

Decanter: A decorative bottle used for wines and spirits.

Deceptive Optic: A term used for heavy bottom glasses which appear to be holding more than their whack.

Dispense: A name given to a small bar in some place away from the main bars, i.e. in a Restaurant, or Club-room. Used for "Waitress Service".

Draught: Beer, or cider, delivered in bulk and dispensed through taps, pumps, or special units—i.e. "drawn".

Dray: The vehicle in which beer is delivered to the house, formerly horse-drawn (from which the word derives) but now usually motorised, although one or two brewers still deliver by horses. The terms "Dray" and "Draymen" are still in common use.

Drip-Can: A container used for catching the drips (and overspill) and placed underneath the beer taps for this purpose.

175

Drip-Cup: A porcelain or metal cup attached by wires under the taps of casks where beer or cider is being drawn from the wood. Also used for the same purpose under spirit bottles dispensing by the "Pearl" optic.

Drop: A name often used for a "nip" or single measure of spirits.

Finings: A thickish, cloudy substance made from isinglass—used to clear beer of all extraneous matter. Not part of the beer itself but merely used in the processing of beer. Sometimes referred to by cellarmen as—

Fishguts: —which name, no doubt, comes from the fact of isinglass being made from the swimming bladder of the sturgeon and other fish. Finings of a different kind are used to clear wines and spirits during the production process.

Flat Beer: Beer which for any number of reasons pours out with no froth on top.

Float: The small cash placed in the till to enable change to be given. Usually a fixed amount left in the till permanently. Insurance companies will cover a "Float" in each till up to £2.50 so there is not too much risk—even if it does happen to be stolen.

Fob: The Brewery name for heavy, frothy overflow, heard when pressure tanks are being filled with draught.

Four-Ale Bar, Four Ale: A name given to the Public-bar—a reminder of the time when ale was sold at fourpence a quart.

Free Trade, Free House: A house "free" to sell any kind of beer, wines or spirits, not being "tied" to any brewer, or firm in any way. (See "Tied" Trade.)

Full Licence: A Licensee holding a licence which entitles him to sell all kinds of alcoholic liquor by retail without restriction. It also includes the licence for "Off-sales".

Funnel, Spile: A funnel with a long thin attachment which is placed through the hole drilled in the shive of casks to enable beer to be worked back slowly below the level of the liquid—thus avoiding much frothing over.

Funnel, Wine: A funnel containing a fine wire gauze, either fixed or loose, used for decanting wines where a sediment is likely to be present.

Head: The froth on top of a glass of beer (also called "the collar").

Head: The top of a cask where the tap-hole is found.

Header (or Spreader): A nozzle placed on the spout of beer pumps to aerate and give a "head" to beer which might otherwise draw up flat.

Jug: Name often given to glass mugs with handles, usually Government-stamped pints and half-pints.

Jug and Bottle: A department set aside in a public-house for the sale of draught beer for "Off" sales in customers' own jugs, or vessels. Not often seen now, and superseded by a proper "Off-licence" department in most modern houses.

Keystone: A wooden bung, partly drilled through, made to fit the tap-hole of a cask, through which the tap is driven. The centre disc is punched through into the cask as it is tapped leaving the outer ring as a wooden sheath to grip the tap and make a tight joint.

Licensee: The person holding the Licence—who is responsible in law for almost everything which goes on in the premises, whether he was present at the time, or not; in some cases he is responsible for that he did not even know of.

Line Glasses: Oversize glasses marked with a line near the top to indicate the level of a pint or half-pint. The extra size is for the purpose of accommodating the "head". The Weights and Measures Department stamp these glasses and it is quite legal to use them.

Liquor: In the Trade water is known as "liquor".

Measuring Taps: There are two kinds of Measuring Taps in general use, the OPTIC PEARL (see note under OPTIC) and "NON-DRIP". The "NON-DRIP" is the modern quick-service method of dispensing spirits which may easily be drawn from two units at a time by merely pushing the rim of the glass upwards against the bar and waiting until the measure is empty—when it will automatically re-fill for the next order.

"Metso": Patent name for a beer pipe cleanser. Others are "Hygex" and "Quatso", etc., etc.

Mug: Same as Jug (which see).

Multiple Tenant: Persons, or firms, holding several Licences are said to be "multiple Licensees" or if in the Tied trade "multiple Tenants".

Music Licence: A house which does not employ *more* than two instrumentalists, or vocalists, does not require a Music Licence —all others do, and many stringent regulations must be complied with concerning public safety, etc., before such a Licence will be granted.

Neat: Spirits straight from the bottle and unmixed with anything else are said to be "neat" or "straight".

Nip: A "single" measure of spirits.

Nip: A small bottle containing approximately one-third of a pint. Strong heavy beers such as Barley Wine, Russian Stout, etc., are bottled in "nip" sizes.

"Off-Licence": A Licence permitting the holder to sell alcoholic drink for consumption off the premises. It is an offence for anyone to permit drinking in "Off-licences" or in the "Off-sales" Department of a Public-house.

Optic: A measuring device which when inserted in a bottle and inverted in a stand permits one measure to be dispensed at a time by moving a lever. It has a bulbous glass front which may be seen emptying and re-filling after each measure has been drawn. Known as the "Optic Pearl".

Overspill: Sound beer which runs over the top of the glass when it is being filled. Not to be confused with Waste (which see).

Palate: Flavour, or taste. "Good palate" means it is agreeable to the taste.

Pewter: The shelving under the counter on which glasses are washed and drained and kept ready for use. Although formerly made of real pewter the modern trend is towards a stainless steel wash-up and draining-board with Formica shelving.

Pilsener, Pilsner: Lager—for a full explanation of which see under this heading in Part 1.

Pilsener: A tall narrow glass used for serving Lager.

Pony: A term used for a glass which will hold one-third of a pint. Used for serving "nip" size bottles and "split" size minerals, etc.

Pourers: There are two kinds of Pourers, which make for ease of pouring (without the "glug-glug-glug" attendant on pouring from an ordinary bottle) but do not measure, and those which pour a stipulated measure each time the bottle is turned upside down. Measuring Pourers, shaped like a ball, fit into the bottle, which may then be carried to the counter for service, instead of the glasses being taken to the Optic or "Non-Drip". *Note*: Forget not to remove them from the counter—otherwise, well! . . . you know! There are two kinds of Measuring Pourers in general use: BOWKER MEASURING POURER, DALEX MEASURING POURER, and there is little to choose between the two except that the Bowker is probably quicker off the mark.

Pressure: A system of delivering beer to the bar dispensing unit under pressure from CO_2 (called Gas) (see *Canister*).

Racked Beer, Racked Bright: This is beer from which every trace of sediment has been removed at the Brewery and which is ready for sale immediately on arrival. It does *not* require to settle, and is called for at Weddings, Parties, etc., where it will be consumed quickly—as it does not keep long.

Racking Cock: Technical name for a beer tap.

Scantling, Stillage, Stillion: All terms used for the heavy wooden beams or racks on which casks are rested, prior to, or after being broached (Wines or Spirits) or Tapped (Beer, Cider, etc.).

Scotches: Wooden wedges used on a scantling to prevent casks from moving.

Shive: A circular wooden disc hammered into the bung-hole of a cask to take the place of one which has been chopped out (for Fining, etc.). Often drilled through for about half the thickness with a small hole into which will fit the spile (which see). The centre hole may be punched through very easily and the spile inserted—without the necessity of drilling through the whole thickness of the shive.

Snug: Usually a small cosy room without a bar, but with waiter service, or served throught a hatch. Not found much in the South.

Spile, Vent Peg: These are conical wooden pegs inserted into the shives of casks (after a hole has been drilled right through the shive to receive them). They have two purposes and there are two kinds—hard and soft. Soft spiles are porous and are used to allow excess gas (condition) to escape from the cask with little or no loss of beer. Hard spiles are non-porous and are used for the opposite reason—i.e. to keep "condition" *in* the beer.

Still: In wines or table waters means the opposite of sparkling.

Stillage, Stillion: See reference under heading of *Scantling*.

Straight: See reference under heading of *Neat*.

Tenant: Many people have only a vague idea of the meaning of the word "Tenant". A "Tenant" is one who enters a house, not as the proprietor, or owner, but as one who holds or possesses it under Agreement from the actual owners—usually a Brewery company. This Agreement constitutes the "tie", for the "Tenant", far from being allowed freedom to purchase and sell any product he wishes, is "tied" by the Agreement to buying certain of his stock (if not all) as directed by his Landlords.

For all this he pays a rent. He will also have paid a sum of money known as "Ingoing". This is to cover the value of certain items left behind by the outgoing Tenant—carpets, furniture, chandeliers, glasses, etc., etc. So for his "Ingoing" and his rent he is allowed to sell any beer under the control of the Brewer Landlord, with Wines and Spirits usually stipulated as well.

If a man says he "bought" a pub for £1,500 you may safely say he didn't buy it—he merely paid that sum as "Ingoing" and leased the premises for a term of years at a rental from the Brewers.

Thimble: Loose metal vessels in which spirits are measured are often known as "thimble" measures.

"Tied" Trade: See *Tenant* above.

Tilt, Tilter: This is an apparatus, either in wood or metal, used for "tilting" casks when on a scantling or stillion, in order to draw off the liquid below the level of the tap. Some are quite elaborate mechanical contrivances, others are merely a triangular wooden frame, notched on one side. These operate by resting against the wall behind the cask—the cask being raised one notch at a time as required. Tilting is a very important part of cellar work in the case of beers not delivered "racked bright". Unless performed with the greatest care the finings and sludge may rise and not settle again, thus spoiling all the beer remaining in the cask.

Trade: The query "Whose Trade are you in?" means "Whose beers do you sell?"

Ullage: A term generally used in the Trade for beer, bottled and draught, which is to be returned to the bottler, or the Brewer, through some fault. (The term applies to Wines, Spirits, Minerals, etc.). It may be that the bottle neck is chipped, the bottle cracked, the crown cork faulty, empty, or the beer itself may be wrong.

Bottles of Wines, Spirits and Liqueurs which have been opened for service are said to be "on ullage". Strictly the word "ullage" means the amount by which a cask falls short of the stipulated quantity.

Waste: The "bottoms" of bottles and beer left behind in customers' glasses, etc. Not suitable for re-sale and quite distinct from genuine "overspill" or "ullage".

Well: The name given to the sink inset in the pewter for washing up.

Wood: The term for casks of all sizes, whether made of timber or metal. Also refers to empty beer cases and crates. The words "Drawn from the Wood" mean that the beer is drawn and served in the bar in view of the customer.

Working: Beer is said to be "working" when it is not "quiet" and the yeast and gas are much in evidence.

APPENDIX TWO

FURTHER TRAINING—BOOKS AND COURSES

If you want to succeed in your career and have mastered the contents of this book, keep up the good work. *Read* as much as you can in the Trade Press and in other books.

Especially recommended are:

INNKEEPING: A Manual for Licensed Victuallers. Published for the Brewers' Society by Barrie & Jenkins Ltd. This is the publican's "bible". A very comprehensive manual containing much that you do not *need* to know as a barman but which you certainly will at a later stage in your career and which you will find interesting to read even now. It is the authorised textbook for the courses run by the Licensed Trade Training and Education Committee (see below).

THE ABC OF THE LICENSING LAWS by the Solicitor to the London Central Board, obtainable from the Licensed Victualler's Central Protection Society of London Ltd., 32 Bedford Square, London, W.C.1.

HOTEL AND CATERING LAW by Frank J. Bull and John D. G. Hooper, published by Barrie & Jenkins Ltd. A general textbook for the layman. More useful for the prospective tenant or club owner than for the barman, but nevertheless a good "background" book for you.

TRAINING COURSES

Try to attend Training Courses. There are excellent courses run specially for the Licensed Trade organised in Centres all over the country by the Licensed Trade Training and Education Committee whose Secretary will be glad to send you details if you write to him at 42 Portman Square, London W1H 0BB. The Secretary of your Licensed Victuallers Association or LTDA will be able to help you here too.

For those able to spare the time, a series of residential courses are run at the Brewers' Society's two residential training centres at Buxton in Derbyshire and Donhead in Wiltshire. Full details may be obtained from the Bookings Secretary, Residential Training Scheme, 42 Portman Square, London W1H 0BB. A course lasting three weeks is recognised by the Training Opportunities Scheme, and applicants may be made at local Job Centres of the Department of Employment.

If you cannot attend one of these courses, see what courses are available locally by calling at the Education Offices (probably at or near the Town Hall). There may well be a general course which will deal with, at any rate, some aspects of your job—though your hours of duty *may* make it difficult for you to attend. The LTTEC courses are organised with the convenience of those working "licensed hours" specially in mind, and students may sit for the examination leading to the Licensed Trade Diploma.